Praise for Philip Goff

GALILEO'S ERR

"*Galileo's Error* is a manifesto for a new generation of philosophers who think we need to revise our view of the physical world to accommodate consciousness. Galileo took the mind out of matter, which was good for the science of matter but not so good for the science of the mind. Philip Goff thinks that to explain consciousness, we have to put the mind back into matter. His ideas are radical, but his arguments are rigorous and the book is a pleasure to read. I recommend it to anyone who wants to come to grips with mystery of consciousness."

—David Chalmers, author of *The Conscious Mind* and professor of philosophy at New York University

"Accessible. . . . Fascinating and somewhat revolutionary."

—*Bowling Green Daily News*

"Provocative, brave, and clearly written, Goff's new book introduces the public to a revolutionary approach to one of the most stubborn of mysteries."

—Lee Smolin, author of *Einstein's Unfinished Revolution* and founding member of the Perimeter Institute for Theoretical Physics

"Goff's elegant book offers a thought-provoking, inspiring picture of the nature of mind. His spirited defense of panpsychism moves well beyond the usual academic discussions, mulling over our place in the larger universe."

—Susan Schneider, author of *Artificial You: AI and the Future of Your Mind*

"*Galileo's Error* is a tour de force. Goff defends his distinctive view of consciousness with verve, wit, and authority. . . . It is hard to imagine a better introduction to current debates about consciousness."
—David Papineau, author of *Thinking about Consciousness* and professor of philosophy of science at King's College London

"*Galileo's Error* is an exciting and provocative book. . . . Goff writes with clarity and passion, and whether you agree or disagree with his conclusions, you will find his book enjoyable, engaging, and deeply thought-provoking."
—Keith Frankish, editor of *Illusionism: As a Theory of Consciousness* and honorary reader in philosophy at the University of Sheffield

"Philip Goff has written an extraordinarily accessible and entertaining book introducing and defending an increasingly popular, if, on the face of it, outlandish, claim: that consciousness is everywhere. . . . There's no better introduction to this fascinating subject."
—Stephen Law, author of *What Am I Doing With My Life?*

"A short and widely accessible full-throated intellectual defense of an ancient view—that mind is much more prevalent than we like to believe in the West, constituting a basic building block of the universe."
—Christof Koch, president and chief scientist of the Allen Institute for Brain Science

Philip Goff

GALILEO'S ERROR

Philip Goff is an Associate Professor of philosophy at Durham University. He is the author of *Consciousness and Fundamental Reality* and has published more than forty academic papers. His writing has also appeared in many newspapers and magazines, including *The Guardian*, *Scientific American*, and *The Times Literary Supplement*, and he has guest-edited an issue of *Philosophy Now*. He lives in Durham, England.

www.philipgoffphilosophy.com

ALSO BY PHILIP GOFF

Consciousness and Fundamental Reality

GALILEO'S ERROR

GALILEO'S ERROR

Foundations for a New Science of Consciousness

Philip Goff

VINTAGE BOOKS
A Division of Penguin Random House LLC
New York

FIRST VINTAGE BOOKS EDITION, OCTOBER 2020

 Published in the United States by Vintage Books, a division of Penguin Random House LLC, New York, and distributed in Canada by Penguin Random House Canada Limited, Toronto. Originally published in hardcover in the United States by Pantheon Books, a division of Penguin Random House LLC, New York, in 2019.

The Library of Congress has cataloged the Pantheon edition as follows:
Name: Goff, Philip, author.
Title: Galileo's error : foundations for a new science of consciousness / Philip Goff.
Description: New York : Pantheon Books, 2019. | Includes bibliographical references (pages 223–228) and index.
Subjects: LCSH: Panpsychism. | Consciousness. | Galilei, Galileo, 1564–1642.
Classification: LCC BD560 .G64 2019 | DDC 128/.2—dc23
LC record available at https://lccn.loc.gov/2018057138

Vintage Books Trade Paperback ISBN: 978-0-525-56477-5
eBook ISBN: 978-1-5247-4797-8

Author photograph © Ian Hobson
Book design by Michael Collica

www.vintagebooks.com

Printed in the United States of America
10 9 8 7 6 5 4 3

For my parents, Marie and Tony, who gave me determination and a passion for the truth

Contents

Acknowledgments

I'd like to thank Emma Bullock for the wonderful illustrations that help bring the thought experiments to life.

I am extremely grateful to David Chalmers and Nigel Warburton for helping me to develop the idea into a popular format. I would especially like to thank Nigel for coming up with the title of the book. The next stage was getting it accepted for publication, and I am indebted to my agent, Max Brockman, for helping me to develop the proposal and for securing a publisher. The only thing left to do then was to write it. My editors, Edward Kastenmeier and Andrew Weber, did a great job drawing my attention to obscurities of the first two drafts and helping me see that things that are obvious when you've been thinking about this stuff for twenty years might not be so obvious to someone coming to the topic for the first time. For really helpful comments on all or part of the text (or on my earlier attempts at a trade book), I would like to thank Luke Roelofs, Hedda Hassel Mørch, David Papineau, David Chalmers, Barry Loewer, Alastair Wilson, Keith Frankish, Galen Strawson, Simon Goff (who provided invaluable last-minute help with some tricky phrasing issues), John Houghton, Clare Goff, Garrett Mindt, Damjan Aleksiev, Marta Santuccio, Ian Hobson,

Tony Goff, Helen Hobson, Rob Hoveman, and Jessy Hawkins. I'd also like to thank Ian Hobson for the photo on the back flap of the jacket.

Most of all, I would like to thank my wife, Emma, for incredible support during the writing process, for countless hours of invaluable discussion of the material, and for rigorous comments on the final manuscript. As always, I'd be lost without you.

GALILEO'S ERROR

I

How Galileo Created the Problem of Consciousness

As you read this page, you are having a visual experience of black letters against a white background. You can probably hear background noises: traffic, distant conversation, or the faint hum of a computer. You may be experiencing strong smells or tastes: the smell of coffee, the taste of mint as you chew a fresh piece of gum. Maybe you feel some emotions: a feeling of excitement, or sadness, or maybe you just feel a bit tired or distracted. If you pay attention, you will notice more subtle kinds of experience: the tactile sensation of the chair against your body, perhaps sensations of itchiness or throbbing in a knee or an arm. These are all forms of *conscious experience.* They are states that characterize your *subjective inner life.* These feelings and experiences make up what it's like to be you.

Consciousness is fundamental to what we are as human beings. This is not to undermine the importance of the body: we are embodied creatures and we relate to one another through and with our bodies. But it is consciousness that defines the identity of the person. Fundamentally we know a person in terms of their feelings and thoughts, and the quirks of their personality. Perhaps one day in the future it will be possible to survive the death of the body by uploading one's mind onto a computer, making it possible to talk to grandma via email long after her body has rotted in the ground. So long as the conscious mind survives, we feel that the person has survived. In contrast, when the conscious mind has gone, for example, in tragic cases of permanent coma, the living body can seem nothing more than a memorial to the person who once was.

Consciousness is also the source of much that is of value in existence. Without consciousness, the universe would still be just as immense and awe-inspiring. But without a conscious mind to appreciate its majesty, is there really any value in the existence of *all that stuff*? It is our experiences that make life worth living: exhilarating pleasures, sweeping emotions, subtle thoughts. Without consciousness none of these things is possible.

As well as being the ground of our identity and a source of great value, consciousness is the only thing we know for certain is real. I can't know for certain that there really is a world out there. Perhaps I'm actually in the Matrix, naked and hairless, encapsulated in a vat of chemicals being fed information about a nonexistent virtual world by computers that are using me as an energy source. I might not even have a body: perhaps the computers disposed of my body long ago and now all that remains of me is a brain wired up to a computer. Or perhaps I am myself a computer, created by humans to think that I am a living human being.

But there is one thing I know for certain: I exist as a conscious being. If I am in the Matrix, the computers might be deceiving me about all kinds of things, but they can't be making me think I'm conscious when I'm not. Perhaps my visual experience of the room around me doesn't correspond to anything real, but I know I am having a visual experience nonetheless. The only thing I have direct access to are my own experiences.

Everything else is known indirectly, believed on the basis of what I experience. All knowledge of reality is mediated through consciousness.

This was the insight of the father of modern philosophy René Descartes (1596–1650), summed up with his famous line "Cogito ergo sum" or "I think, therefore I am." This phrasing is potentially misleading. Descartes wasn't saying that he exists *because* he thinks. (Hence the following old joke doesn't really work: Descartes goes into a bar. Barman says, "Want a beer?" Descartes replies "I think not" and disappears.) Descartes's point is about *knowledge:* he knows for certain that he thinks—or more broadly that he is a conscious being—and in knowing this he thereby knows that he exists. The certain knowledge of one's existence as a conscious being was, for Descartes, the starting point of all knowledge.

Nothing is more certain than consciousness, and yet nothing is harder to incorporate into our scientific picture of the world. We now know a great deal about the brain, much of it discovered in the last eighty years. We understand how neurons—the basic cells of the brain—work in terms of their underlying chemistry. We know the function of many regions of the brain, in terms of processing information and negotiating sensory inputs and behavioral outputs. But none of this has shed any light on how the brain produces consciousness.

Some people dismiss this as simply reflecting the fact that the physical science of the brain—neuroscience—has a long way to go. But if explaining consciousness is a work in progress, one might reasonably expect neuroscience to have yielded a partial explanation of consciousness, accounting for some human experiences but leaving trickier cases unexplained. The reality is that, for all its virtues, neuroscience has thus far failed to provide even the beginnings of an explanation.

This is all the more extraordinary when we contrast it with the great progress science has made in explaining other phenomena. The scientific story of water or gasoline explains the observable characteristics of these substances. We get a satisfying account of why, for example, water boils at 100 degrees centigrade or of why gasoline is flammable. Our scientific understanding of genes continues to provide ever greater insight into how certain traits are passed on from generation to generation. Astrophysics is able to explain how stars and planets are formed. In all of these cases we find satisfying explanations. And yet our increased understanding of the electrochemical processes of the brain has failed to yield insight into how those processes give rise to a subjective inner world.

WE'LL GET THERE IN THE END . . .

Physical science has a dismal track record in explaining consciousness. But the track record of physical science in explaining pretty much everything else is impressive. Many scientists and philosophers take this to be good evidence that, in spite of current disappointments, neuroscience will one day crack the mystery of consciousness.

The neuroscientist Anil Seth makes an analogy to life.[1] It used to be thought that life was an inherently mysterious phenomenon, which could be explained only via the postulation of mysterious nonphysical "vital forces." Few people these days take this view, known as "vitalism," seriously. According to Seth, this was not because some philosophers solved "the problem of life." We moved on from the days of vitalism because biochemists, instead of dwelling on the mystery, got on with the job of explaining the properties of living systems—metabolism, homeostasis, reproduction, etc.—in terms of underlying mechanisms, and eventually the sense of mystery dissipated.

Seth urges us to take a similar approach to consciousness. It's what we might call the "Get out of the armchair and into the lab" approach. Seth recommends that instead of dwelling on why consciousness exists in the first place, we should rather focus on what he calls the "real" problem of consciousness: the challenge of mapping correlations between what goes on in the brain and what is experienced by the person. Seth predicts that, just as in the case of life, the sense of mystery will eventually go away and scientists of the future will wonder what the philosophers were worrying about.

The trouble with picking examples from the history of science is that there are always other examples that prove the opposite point. Seth focuses on the scientific challenge of explaining life *as it currently exists.* But consider instead the riddle of explaining the *historical origins* of complex life. Before Darwin, it was a mystery where complex, self-replicating organisms came from. The nineteenth-century philosopher William Paley argued that the only plausible hypothesis was that they were created by an intelligent designer, that is to say, by God.[2] Paley argued for this with the following analogy. Imagine you're walking along a beach and you come across a watch lying on the ground. It would be crazy to suppose that something so complex had come about by a chance, random process, and so you would naturally assume that somebody had designed it. Similarly, argued Paley, given the great complexity of living organisms, we should suppose that they too were designed rather than that they came about by chance.

You might think at first this case is quite similar to Seth's. In both cases what was once taken to be the product of mysterious nonphysical interventions in the natural world came to have a scientific explanation. But there is an important difference. As Seth says, the problem of explaining life as it currently exists—the problem which led some in the nineteenth century to postulate nonphysical vital forces—was not solved by some great insight that pointed toward a solution; it just eventually stopped seeming like a real problem in the first place. But the problem of explaining the historical origins of life *was* solved by just such an insight. Darwin didn't just say, "Stop wasting time worrying about where life came from and get on with more serious scientific questions"; rather he came up with the principle of natural selection to explain how complex life emerges. Darwin agreed with Paley that the emergence of complex organisms can't possibly have happened by chance, but rather than appeal to God he postulated the "blind watchmaker"—to use Richard Dawkins's memorable phrase—of natural selection in order to explain it.[3]

Coming back to the problem of consciousness, it seems that it could go either way. Perhaps Seth is right that as we learn more about the brain we will eventually stop worrying where consciousness came from (although there is no sign of this happening yet). But it could equally be that the "Darwin of consciousness" will come along and solve the problem of consciousness in a satisfying way. In opposition to Seth, I will try to show not only that there is good reason for taking the problem of consciousness seriously, but also that there are already the makings of a theoretical framework that could bring about progress.

One of the most vociferous proponents of the "Get out of the armchair and into the lab" approach is the neurophilosopher Patricia Churchland. Imagine, if you can, a fearsome firebrand preacher but for the cause of neuroscience rather than religion, and you're probably on the right track. Patricia Churchland

and her husband, Paul, achieved academic fame in the 1990s for defending a radical position known as "eliminative materialism." Paul and Patricia argued that we should not be trying to explain the mind but rather rejecting its very existence. Like fairies and magic, science has shown that mental phenomena simply do not exist.[4]

Here's an analogy. What people used to think of as demon possession we now know to be epilepsy. But we don't say, "Great, we now have a scientific explanation of demon possession." Instead, the scientific explanation has displaced the demon possession explanation, proving beyond reasonable doubt that demon possession does not exist (or at least it's not what's going on in standard epilepsy cases). Similarly, the Churchlands claimed that our old-fashioned explanations of human behavior in terms of things like "thought," "desire," "hope," "love" were becoming outdated. They looked forward to a day when we would drop such an antiquated vocabulary altogether and talk about human behavior in terms of its real causes: electrochemical processes in the brain. (One can't help wondering whether the Churchlands' early courtship involved poetry expressing the strength of their neuronal activations for each other. . . .)

I am very much open to the idea that scientific progress can show that many of our commonsense ways of thinking about the world are wrong. Modern science has revealed to us that objects we think of as solid are in fact mostly empty space, given the immense distances between the nucleus at the center of the atom and the electrons that orbit it. Einstein's theory of relativity entails that our commonsense notion of absolute time is an illusion. And, as we shall explore in the next chapter, quantum mechanics has undermined many of our commonsense ways of thinking about matter. However, there is a limit to this. One thing that science could never show is that consciousness does not exist.

Imagine reading the following story in the *New Scientist* magazine:

> Scientists Discover That
> Consciousness Is an Illusion
>
> For millions of years humans have believed that they have feelings and experiences. In a shocking development, neuroscientists at the California Institute of Technology have discovered that nobody has ever felt or experienced anything. Dr. Ivor Cutler is team leader on the project:
>
> "We are not disputing any of the commonly known facts about human behaviour; nobody could deny, for example, that people scream when their body is damaged. But the popular belief that bodily damage is accompanied by a *feeling of pain* is in fact an illusion. Feelings are no more real than the Loch Ness Monster."
>
> Lawyers have been discussing the possible impact of this new discovery on human rights legislation.

We would never, and should never, accept such claims. And this is because our scientific knowledge of the world is itself mediated through conscious experience. We are able to perform observations and experiments only because we have conscious experience of the world around us. In this sense, scientific knowledge is dependent on the reality of consciousness. Science could no more prove that consciousness does not exist than astronomy could prove that there are no telescopes.

The basic reality of consciousness is a datum in its own right. Science can tell us all sorts of weird and wacky things about the world: that time doesn't flow, that there are no solid objects, that we're not really free in the way we think we are. But science can't tell us that we don't feel pain or see red. The reality of one's feelings and experiences is immediately known in such a way that their existence cannot seriously be doubted.

WHAT IS "SCIENCE" ANYWAY

There is a worry that the "Get out of the armchair and into the lab" approach can lead to an oversimplistic conception of what science is, as though science were simply a matter of setting up experiments and then recording the data. In fact, certain crucial scientific developments have involved radically *reimagining nature,* dreaming up possibilities—perhaps from the comfort of an armchair—that nobody had previously entertained.

Here are just a few of the most important ways in which great scientists have reimagined nature:

Uniting Heaven and Earth

Popular myth tells us that Newton was the first person to realize that apples fall to the ground. Of course, he wasn't. But he was the first person to entertain the idea that what makes apples fall to the ground is the same thing that keeps the moon in orbit around the earth. It had not previously occurred to anyone that a single force might be responsible for both of these phenomena. What now seems to us so natural was at the time an inspired leap of the imagination.

Uniting Space and Time

Before the twentieth century, scientists had taken it for granted that space and time are different things. Indeed, time and space do seem to have very different characteristics: time flows from past to future, while space seems to be "all there" at once. It was thus a radical reimagining of nature when Hermann Minkowski, in his mathematical interpretation of Einstein's special theory of relativity, dispensed with "space" and "time" as distinct entities, and replaced them with a single entity: spacetime. As Minkowski boldly put it, "Henceforth space by itself, and time by itself, are doomed to fade away into mere shadows, and only a kind of union of the two will preserve an independent reality."[5]

Uniting Gravity and Inertia

Just as nobody before Newton had dreamed of identifying the force which pulls apples to the ground with the force which keeps the moon in orbit, so nobody before Einstein had dreamed of identifying gravitational force (the force which pulls apples to the ground and keeps the moon in orbit) with inertial force (the force which pushes you back in your car seat when you accelerate). And that was not all: in Einstein's baffling reimagining of nature, gravitational force is the result of curvature in the fabric of spacetime. We can only be in awe of the imagination that was able to dream up such a picture of the world!

Of course, all of these novel reimaginings of nature were subsequently tested with observation and experiment in order to work out whether we have any reason to think that they are true.

Nonetheless, the point remains that many important moments in scientific progress involved dreaming up new possibilities, conjuring up in the imagination new ways of thinking about the universe. This is overlooked if we have a conception of science that is too focused on experiment and observation. In the years when he was developing special relativity, Einstein wasn't busy conducting experiments; rather he was staring into space wondering what would happen if you rode on a beam of light.

When we neglect the role of deep thought in science, we close off options. It is quite possible that progress on our scientific understanding of consciousness will be made not only through observation of the brain—important as that is—but also through radically reimagining mind and brain, dreaming up new possibilities not so far entertained by our theories. In chapter 4 of this book, we will explore one proposal for doing precisely this.

There is a further problem with the approach of Seth and the Churchlands. They are assuming not only that consciousness can be explained scientifically, but that it can be explained by the scientific method *as we currently envisage it*. However, there is good reason to think that explaining consciousness will require a change in our understanding of what science is, a change as fundamental and wide-ranging as that which occurred at the start of the scientific revolution. This is because, as we will discover in the next section, the scientific revolution itself was premised on *putting consciousness outside of the domain of scientific inquiry*. If we ever want to solve the problem of consciousness, we will need to find a way of putting it back.

THE PHILOSOPHICAL FOUNDATIONS OF SCIENCE

While Descartes was the father of modern *philosophy,* Galileo Galilei (1564–1642) is standardly thought of as the father of modern science. His formulation of the mathematical laws of nature—the precursor of Newton's (1643–1727) laws of motion and universal gravitation—and his defense of the Copernican view that the earth and planets revolve around the sun—for which he was persecuted by the Church—laid the foundations for the scientific revolution. What is perhaps less remarked upon is that Galileo was also one of the greatest *philosophers* who ever lived. There are at least two respects in which his contribution to the scientific revolution was a philosophical, rather than a scientific, achievement:

The scientific revolution marked the overturning of the *Aristotelian orthodoxy;* that is to say, the widespread acceptance of the worldview of the ancient Greek philosopher Aristotle (384–322 BCE). Aristotle's worldview was complex and multifaceted, but there are two core features in particular that were rejected in the scientific revolution:

- Aristotle adhered to the *Ptolemaic* view of the universe, according to which the earth was in the center of the universe, with the stars and planets orbiting around it.
- Aristotle's theory was *teleological:* inanimate objects had goals built into them that explained their movement. For example, matter falls to the ground because it aims to get back to its natural home in

> the center of the universe, while fire rises because its natural home is in the heavens.

It is popularly assumed that Aristotle's view was shown to be false through the new experimental method, particularly through observing the heavens with the aid of telescopes. Of course, there is a great deal of truth in this. However, it is crucial to note that Galileo managed to reject one crucial plank of Aristotle's theory of the physical universe not through observation or experiment, but through pure philosophical argument. Galileo proved that Aristotle's view that heavy objects fall to the ground faster than lighter ones—a doctrine of common sense that had been believed for thousands of years—was logically incoherent. We will discuss Galileo's argument in more detail in chapter 3.

Moreover, Galileo's most fundamental reimagining of nature—more fundamental than his embracing of the Copernican model of the universe—was never justified by observation or experiments. It was, and remains, a piece of philosophical speculation. And it is this philosophical speculation—which to this day underlies our scientific picture of the universe—that is to blame for the contemporary problem of consciousness. Let me explain.

One of Galileo's most significant contributions to the scientific revolution was his radical declaration of 1623 that *mathematics* is to be the language of science:

> Philosophy [by which Galileo meant "natural philosophy," i.e., what we now call "natural science"] is written in this grand book, the universe, which stands continually open to our gaze, but it cannot be understood unless one first

> learns to comprehend the language and read the letters in which it is composed. It is written in the language of mathematics, and its characters are triangles, circles, and other geometrical figures, without which it is humanly impossible to understand a single word of it; without these, one wanders about in a dark labyrinth.[6]

Why had previous thinkers not framed their theories of nature in mathematical language? The problem was that before Galileo philosophers took the world to be full of what philosophers call *sensory qualities,* things like colors, smells, tastes, and sounds. And it's hard to see how sensory *qualities* could be captured in the purely *quantitative* language of mathematics. How could an equation ever explain to someone what it's like to see red, or to taste paprika? How could an abstract mathematical description convey the sweet smell of flowers?

But if mathematics cannot capture the sensory qualities of matter—the redness of a tomato, the spiciness of paprika, the smell of flowers—then mathematics will be unable to completely describe nature, for it will miss out on the sensory qualities. This posed a severe challenge for Galileo's hope that the "book of the universe" might be written in an entirely mathematical language.

Galileo solved this problem with a radical reimagining of the material world. In this reimagining material objects don't really have sensory qualities. Paprika isn't really spicy, flowers don't really smell of anything, objects aren't really colored. In Galileo's reimagined world, material objects have only the following characteristics:

- Size
- Shape

- Location
- Motion

Hence, for Galileo, the lemon I see in front of me isn't really yellow, and it doesn't really have a citrus smell and a sour taste. In reality the lemon is simply a thing which has a certain size, shape, and location. Of course, the lemon has parts, and there will be a great deal of complexity involved in the arrangement of and the relationship between these parts. But all of that complexity can, according to Galileo, be wholly characterized in terms of the sparse characteristics mentioned above: size, shape, location, and movement.

What is so special about the characteristics of size, shape, location, and movement? The crucial point is that these characteristics can be captured in mathematics. Galileo did not believe that you could convey in mathematical language the yellow color or the sour taste of the lemon, but he realized that you could use a geometrical description to convey its size and shape. And it is possible in principle to construct a mathematical model to describe the motion of, and the relationships between, the lemon's atoms and subatomic parts. Thus, by stripping the world of its sensory qualities (color, smell, taste, sound), and leaving only the minimal characteristics of size, shape, location, and motion, Galileo had—for the first time in history—created a material world which could be entirely described in mathematical language.*

But what of the sensory qualities? If the yellowness, the citrus smell, and sour taste aren't really in the lemon, then where are

* Galileo's attempt to mathematize nature was not without precedent. We find similar moves, for example, in Plato's *Timaeus*. However, this was the first time a mathematical theory of nature had achieved widespread acceptance.

they? Galileo had an answer for this too: the soul.* For Galileo, the lemon itself isn't really yellow; rather yellowness exists in the soul of the person perceiving the lemon. Likewise, neither the sour taste nor the citrus smell are really in the lemon; rather they're in the soul of the person tasting or smelling the lemon. Just as beauty exists only in the eye of the beholder, so colors, smells, tastes, and sounds exist only in the conscious soul of a human being as she experiences the world. In other words, Galileo transformed the sensory qualities from features of things in the world—such as lemons—into forms of consciousness in the souls of human beings.

Consider the age-old philosophical conundrum, "If a tree falls in a forest, and there's nobody there to hear it, does it make a sound?" In Galileo's reimagining of the world, the answer is a clear and resounding *no*. The falling tree produces vibrations in the air, vibrations which have the mathematical characteristics of size, shape, location, and motion. But it is only when there is a soul around to react to these vibrations that a sound comes into existence. Sound for Galileo is not a feature of the material world, but a form of consciousness existing only in the conscious soul of a human being.

Thus, Galileo's universe was divided up into two radically different kinds of entity. On the one hand, there are material objects, which have only the mathematical characteristics of size, shape, location, and motion. On the other hand, there are souls enjoying a rich variety of forms of sensory consciousness in response to the world. And the benefit of this picture of the world was that the material world with its minimal characteris-

* In contrast to Descartes, Galileo followed Aristotle in conceiving of the soul as essentially embodied. Nonetheless, it is clear that he took the soul to be incorporeal and outside of the domain of "natural philosophy," i.e., physical science.

tics could be entirely captured in the language of mathematics. This was the birth of mathematical physics.

By appreciating this radical division, we can see that Galileo certainly did not take physical science (or "natural philosophy" as he called it) to be a complete account of the world. Physical science, for Galileo, was limited to describing only the material world: its purely quantitative vocabulary meant that it was unable to capture the sensory qualities that reside in the soul. Galileo is the father of physical science, but he only ever intended it to provide us with a partial description of reality.

One might question how much these facts about the origins of physical science bear on the contemporary scientifically informed understanding of the universe. Just because Galileo thought that physical science could not explain the sensory qualities, it doesn't mean that he was right. Perhaps the scientific method Galileo brought into existence is more powerful than he could ever have imagined.

It is certainly true that Galileo might have been wrong. But these reflections on the origins of physical science do suggest a response to the arguments of Seth and many others that appeal to the incredible track record of science to support the idea that physical science will one day explain consciousness. Physical science has indeed been extraordinarily successful, but we need to bear in mind that its success began when Galileo took the sensory qualities (sounds, smells, tastes, odors) out of its domain of inquiry: by reimagining them as forms of consciousness residing in the incorporeal soul. The fact that physical science has been extremely successful when it ignores the sensory qualities gives us no reason to think that it will be similarly successful if and when it turns its attention to the sensory qualities themselves, this time as forms of consciousness.

Consider the following analogy. A typical academic job (at

least in a U.K. university) has three quite different components: teaching, research, and administration. The skills which make one good at research are quite different from the skills that make one good at teaching, which are in turn quite different from the skills that make one good at administration. In my first term as a philosophy lecturer, my head of department was kind enough to simplify the job for me by allowing me to focus only on teaching and research. It turned out that when I focused only on these aspects of the job, I did a pretty good job (even if I do say so myself). But that in itself was no reason to think that when I eventually had to turn my attention to administration I would do just as well. Sadly, I was hopeless.

Analogously, the success of physical science in the last five hundred years is due to the fact that Galileo narrowed its scope of inquiry. Just as my head of department said to me, "Don't bother for now with administration," so Galileo said to physical scientists, "Don't bother for the moment with the sensory qualities." The argument from "Physical science has been extremely successful" to "Physical science will one day explain the sensory qualities of consciousness" is not supported by the history of science.

Let me repeat for the sake of clarity: I'm not saying that this proves that physical science cannot explain consciousness. But it does undermine arguments that try to show that it inevitably will.

GALILEO'S ERROR

Popular science programs often tell the following story: For thousands of years philosophers tried to work out what reality was like just by sitting around thinking about it, and then one day Galileo came along and said, "I know, let's find out what the world is like by *observing* it." While the development of a new experimental method was crucial, an exclusive focus on this ignores the *philosophical* underpinnings of our current conception of natural science. Galileo the philosopher created physical science by setting the sensory qualities outside of its domain of inquiry and placing them in the conscious mind. This was a great success, as it allowed what remained to be captured in the quantitative language of mathematics.

However, those sensory qualities have come back to bite us, as we now seek a scientific explanation not only of the inanimate world but also of the conscious mind. And we cannot divorce the subjective inner world of consciousness from the sensory qualities which populate it: the colors, smells, tastes, and sounds that characterize every second of our waking experience. An "explanation" of consciousness that is unable to account for these sensory qualities would in fact be nothing of the sort. If Galileo traveled in time to the present day to hear that we are having difficulty giving a physical explanation of consciousness, he would most likely respond, "Of course you are, I designed physical science to deal with *quantities* not *qualities*!"

Physical science is a wonderful thing. And it was only possible because Galileo taught us how to think of matter mathematically. However, Galileo's philosophy of nature has also bequeathed us deep difficulties. So long as we follow Galileo in thinking (A) that natural science is essentially *quantitative* and (B) that the *qualitative* cannot be explained in terms of the *quantitative*, then consciousness, as an essentially qualitative phenomenon, will be forever locked out of the arena of scientific understanding. Galileo's error was to commit us to a theory of nature which entailed that consciousness was essentially and inevitably mysterious. In other words, Galileo created the problem of consciousness.

How can we correct this error? In the forthcoming chapters we will consider three possibilities:

First Option: Naturalistic Dualism

Proponents of the first option accept Galileo's *dualism,* that is to say, his division of nature into two distinct categories: physical objects with their mathematical properties and incorporeal minds with consciousness. Immaterial minds are normally taken to be beyond scientific understanding. However, the *naturalistic* dualist denies that the conscious mind is something magical or mysterious, instead taking it to be part of the natural order. Whereas Galileo set the soul outside of the domain of natural science, the naturalistic dualist wants to expand science in such a way as to include nonphysical minds. Naturalistic dualism will be the focus of the next chapter.

Second Option: Materialism

Materialists are grateful to Galileo for creating physical science, but respectfully disagree with his conviction that consciousness is a real phenomenon that resists physical explanation. This is the position of Seth and the Churchlands discussed above. Radical materialists argue that consciousness is an illusion. More moderate materialists hope that we will one day be able to explain the subjective inner world of consciousness in terms of the chemistry of the brain. In either case, materialism offers a *conservative* correction of Galileo's error, one that does not require a new paradigm of scientific explanation. Materialism will be the subject of chapter 3.

Third Option: Panpsychism

Recent thought about consciousness has been dominated by the above two options. The dualists argue that there can never be a physical explanation of consciousness, while the materialists retort that the soul can never be part of science. It is hard to see how this perennial debate could ever be resolved in favor of one or the other option. But there is a theory which concedes that there is an element of truth in each of these arguments: panpsychism. Panpsychists believe that consciousness is a fundamental and ubiquitous feature of the physical world. An increasing number of philosophers and even some neuroscientists are coming around to the idea that it may be our best hope for solving the problem of consciousness. In chapter 4 I will explain why.

It is too early to say for sure which of these solutions, if any, will solve the problem of consciousness. But you cannot solve a problem unless you have a deep understanding of what exactly the problem is. The problem of consciousness began when Galileo decided that science was not in the business of dealing with consciousness. To solve the problem, we must somehow find a way of making consciousness, once again, the business of science.

2

Is There a Ghost in the Machine?

Consider your best friend; let's call her "Susan" for the sake of discussion. Imagine Susan is sitting in a chair in front of you with her back to you. You are standing behind looking down at the top of her head. Now imagine that the top of Susan's head is surgically removed so you can stare down into the soggy gray brain matter within. The human brain is an extraordinarily complex organ, involving almost a hundred billion neurons each directly connected with ten thousand others, yielding some ten trillion nerve connections. But, you might reasonably wonder, where in those trillions of neural connections is *Susan:* her hopes and fears, pleasures and pains, the indefinable essence of her personality?

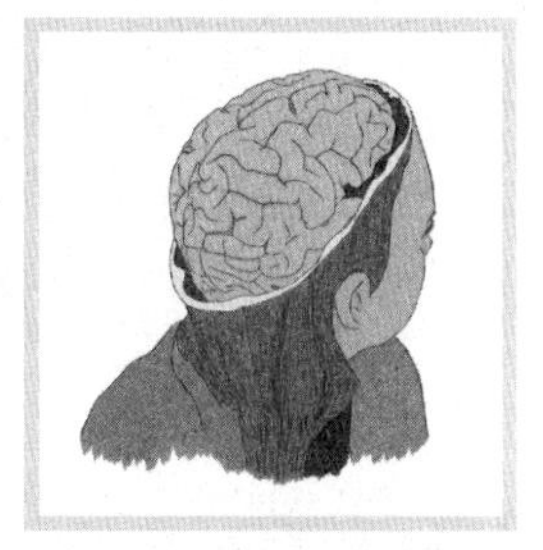

Such reflections help to make vivid the core of the problem of consciousness. Our fundamental conception of each other is as *conscious beings,* that is to say, creatures with feelings, experiences, and emotions. And yet feelings and emotions don't seem to show up in the scientific story of the body and brain. How do these seemingly radically different stories of the human person fit together?

Historically, the most popular solution to the problem of consciousness is ***dualism.*** According to dualism, reality is made up of two very different kinds of thing: immaterial minds on the one hand and physical things on the other. Dualism defines minds and physical things in opposition to each other. Minds are nonphysical: they have no shape, size, or weight; they cannot be observed with any of the five senses. What characteristics *do* minds have? According to dualism, minds are the bearers of consciousness: it is the mind, not the brain, that thinks and feels. Physical entities, in contrast, have no mental characteristics. They have only the characteristics that we can observe or that we learn about in physical science: size, shape, mass, etc.*

Return again to gazing down at Susan's brain peeking out of the top of her head. If dualism is true, what you are now looking at is not Susan. Strictly speaking, Susan is an invisible mind; what you are looking at is merely the body and brain that Susan uses to negotiate the world.† For the dualist, the relationship between Susan and her physical body is a bit like the relationship between a drone pilot and his drone. Just as

* Some contemporary dualists subscribe to ***property dualism.*** This is the view that although there are no nonphysical *things* (there are only physical things like bodies and brains), some physical things (e.g., brains) have both physical and nonphysical characteristics (or "properties"). For the sake of simplicity, I will ignore this more nuanced form of dualism. I think the issues we will discuss in this chapter look the same regardless of whether one opts for property dualism or for the more traditional ***substance dualism*** (the difference being that the latter involves a commitment to nonphysical minds as well as nonphysical consciousness). More detail can be found in my academic book, *Consciousness and Fundamental Reality.*

† I am assuming here that the dualist will give the same answer to the questions "What is Susan?" and "What is Susan's mind?" In principle, this might not be the case; the philosopher John Locke (1632–1704), for example, was inclined to dualism about the mind but did not equate the person with her soul. However, I will ignore this complication as it is not important for the issues we are discussing here.

the drone pilot controls the drone and receives information about the world from it, so Susan controls (to an extent) her body and receives information from its eyes and ears. But Susan is not the same thing as her body or brain, and could perhaps exist without them. According to dualism, a human being is a kind of *composite entity:* a combination of a physical body and an immaterial mind.

Dualism does solve a lot of problems. Much effort in the science of consciousness is spent trying to explain consciousness in terms of the electrochemical processes of the brain. This project has not thus far had a great deal of success despite rapid progress in our neuroscientific understanding of the brain. The dualist has a simple explanation of this: mind and brain are entirely distinct and independently existing things. Feelings, emotions, and experiences reside not in the brain but in the nonphysical mind. Minds and bodies certainly seem to be very different things. The dualist takes this appearance at face value.

Whether or not it's true, dualism is a very natural way to think about ourselves. Most of the cultures and religions we know about have embraced dualism of some form or other. And even some of the most vigorous opponents of dualism concede that in ordinary life they cannot help thinking of their minds as distinct from their bodies. The psychologist Paul Bloom has argued that dualist thought is hardwired into us and that from an early age children start to categorize "mental things" as distinct from "physical things."[1]

Just because a view comes to us naturally or is hardwired, it doesn't mean it's true, but it doesn't mean it's false either. In thinking about dualism, as with all theories of consciousness, we should consider the evidence and the arguments without prejudice.

THE INTERACTION PROBLEM

As all philosophy undergraduates learn, the most famous dualist in Western philosophy was René Descartes. In the last chapter we introduced Descartes's idea that the mind is better known than the body: you don't know for certain that you have a body or a brain, but you do know for certain that you exist as a conscious being. Descartes inferred from this that the mind and body must be distinct things.

When philosophy undergraduates learn about the dualism of Descartes, they also learn that some of the most serious challenges to dualism arise from the difficulty of understanding how mind and body *interact*. Dualists do not want to deny that the mind can affect the body, for example, when the conscious mind feels pain it causes the body to cry out. Nor do they want to deny that the body affects the mind, e.g., images hitting the retina of the eye cause the conscious mind to have visual experiences. But how is this possible? Above we compared the relationship between Susan's mind and her body (according to dualism) to the relationship between a drone pilot and his drone. But many would argue that there is an important dis-analogy between the two cases: we have a good understanding of how the pilot communicates with the drone via the transmission of radio waves; we have zero grasp of how a nonphysical mind could interact with a physical brain. This challenge was pressed on Descartes by Princess Elisabeth of Bohemia in a lengthy and fascinating correspondence.

In fact, this challenge is less straightforward than is often presented in undergraduate philosophy courses. It is true that the dualist cannot explain the causal connection between the mind and brain, but this is arguably part of a more general limitation of human knowledge. While Descartes struggled to make sense of mind-body interaction, a century later the great Scottish philosopher David Hume (1711–1776) pointed out we cannot explain *any* of the fundamental causal relationships of the universe.[2] If this is true, then the "failure" of the dualist is just one instance of a more general failure of our scientific understanding.

To make Hume's point clear, consider Newton's theory of gravity. Newton's theory didn't really *explain* gravity; rather it proposed a mathematical law that describes the gravitational attraction between all material entities. According to Newton's law, all material entities attract each other with a force that is proportional to their mass and inversely proportional to the distance between them. Newton didn't explain *why* objects exert this force. And he admitted it: concerning the underlying mechanism that generates gravity, Newton famously said "Hypotheses non fingo" or "I frame no hypotheses."

Nearly three hundred years later Einstein did propose a deeper explanation of gravity, in terms of the idea that matter *curves spacetime.* This resulting curvature then impacts on the paths of material bodies: things have—according to Einstein's theory—an innate tendency to follow the shortest paths through spacetime, and what ends up being the shortest path is determined by how spacetime is curved. However, just as Newton didn't provide an explanation of why objects exert gravitational force, so Einstein didn't provide an explanation either of why matter curves spacetime or of why matter follows the shortest path through spacetime. When it comes to the basic causal workings of the universe, scientists provide mathematical laws which describe with great accuracy *how* matter behaves, but they provide no explanation of *why* matter behaves in that way.

Return to our drone analogy. I said above that we have a

good understanding of how the pilot communicates with the drone via the transmission of radio waves. This is true in the sense that we know the laws of physics that govern electromagnetic waves, of which radio waves are one form. But we have no explanation as to why nature behaves in accordance with those laws. As basic laws of physics, they simply have to be taken for granted. So digging deeper, we discover that standard physical interactions are just as brute and mysterious as the interactions between minds and bodies postulated by the dualist.

To be clear, I'm not saying that scientists *ought* to explain why the fundamental laws of nature obtain. The point is: if it's okay for the physicist to postulate basic and unexplained laws governing the causal interactions of matter, why isn't it also okay for the dualist to postulate basic and unexplained laws governing the causal interactions of mind and brain?

NATURALISTIC DUALISM

While Descartes is the most famous historical proponent of dualism, the Australian philosopher David Chalmers is probably the most famous dualist alive today. Chalmers shot to philosophical fame in the 1990s for his pioneering work on the philosophy of consciousness. While still only in his twenties, he was the rock star philosopher with long hair and leather jacket, looking more like a heavy metal guitarist than an academic. Chalmers changed the science of consciousness forever with a simple three-word phrase that has become an essential part of the academic discourse on consciousness: The Hard Problem.[3]

Why did this simple phrase have such an impact? For much of the twentieth century consciousness was a taboo topic, seen as a somewhat mystical notion not suitable for "proper" scientific study. The high point of this consciousness-phobia was the behaviorism of the 1920s and 1930s, pioneered by the psychologists John B. Watson and B. F. Skinner. Psychologists in the nineteenth century had investigated the mind by introspecting their own conscious experience. Watson and Skinner objected to this method as nonscientific: people's private experiences can't be rigorously examined in the lab. The behaviorist creed was that the only proper subject matter for scientific psychology was observable behavior.

Toward the end of the twentieth century, it was starting to become acceptable again to use the word "consciousness" and even to want to explain it scientifically. However, Chalmers argued that most of the "explanations of consciousness" being proposed were nothing of the sort. These theories claimed to be explaining consciousness, but in reality they focused on some behavioral phenomenon closely associated with consciousness. They took, for example, the human ability to report verbally on internal states, labeled that "consciousness," and then claimed that in explaining this behavioral capacity they had thereby explained consciousness.

Chalmers categorized theories explaining the behavioral manifestations of consciousness as dealing with "the easy problems" of consciousness (although, as he concedes, even the easy problems are extremely hard!). The "hard problem" of consciousness, in contrast, was the challenge of explaining why the activity of the brain gives rise to *experience:* feelings, emotions, sensations, the subjective inner world each of us knows in her or his own case. As we discussed in the last chapter, neuroscience struggles to explain why subjective experience exists at all. With his simple three-word phrase, Chalmers swept away decades of evasion and forced us to confront the real mystery head-on.

Chalmers's own favored solution to the hard problem of

consciousness is an approach he terms "naturalistic dualism." Historically most dualists have identified the mind with the soul, taking it to be beyond the realm of scientific explanation. In Eastern religion, the soul exists in an infinite cycle of reincarnation driven by the law of karma, which ensures that the quality of one's rebirth is determined by the moral character of one's actions in previous lives. In Western religion, the soul is miraculously created by God at the moment of conception and taken after death either to eternal bliss or to eternal torment. Clearly these divine or karmic actions are not the kind of thing we could hope to capture in an equation.

Naturalistic dualists, in contrast, seek to bring the nonphysical mind into the realm of serious scientific study. I once asked Chalmers if he has any spiritual beliefs or religious commitments. He answered, "Only that the universe is cool." Another leading naturalistic dualist, the German-born Swiss philosopher Martine Nida-Rümelin, is positively angry at the idea that dualism should be mixed up with the myths and vagaries of religious belief. She once remarked to me at a philosophy conference dinner, "The whole concept of faith is fundamentally irrational. It asks you to believe without evidence!"

Nida-Rümelin entirely rejects the idea that we should think of the immaterial mind as something magical and mystical. It is simply a part of nature, as natural as an electron or a planet. And like electrons or planets, minds are—according to naturalistic dualism—governed by natural law. This is not to say that minds are governed by the laws of *physics;* after all they are not physical. Rather the naturalistic dualist postulates special *psycho-physical laws:* basic principles of nature—as fundamental as the laws of gravity or electromagnetism—that govern the interactions between the nonphysical mind and the physical world.

What would be an example of a psycho-physical law? And

how on earth can we know what the psycho-physical laws are? According to the naturalistic dualist, this is a scientific question and so we should look to science for the answer. In fact, the naturalistic dualist employs pretty much the same scientific method as the materialist. Contrary to common assumption, the data of neuroscience is neutral on the true explanation of consciousness and can be interpreted in either a dualist or a nondualist framework. Let us explore this point in some more detail.

To demonstrate the neutrality of neuroscience, consider the *Integrated Information Theory*—IIT for short—a leading contender in the contemporary neuroscientific study of consciousness. IIT is the brainchild of the neuroscientist Giulio Tononi, and is also vigorously defended by Christof Koch, who formerly collaborated on the topic of consciousness with the Nobel Laureate and co-discoverer of DNA Francis Crick. Tononi has formulated a mathematically precise way of defining the quantity of *integrated information* in a physical system. Very roughly, his proposal is that this measure of integrated information, which he labels "ϕ," is also a measure of consciousness.

IIT is able to explain much of the empirical data about consciousness. We know, for example, that consciousness is not correlated simply with complexity. If we compare the cerebellum—a region of the hindbrain that plays an important role in motor control—and the cerebrum—the large, uppermost region of the central nervous system—we find that the former has many more neurons than the latter: 69 billion of the brain's total 86 billion. And yet it is parts of the cerebrum, such as the posterior cortex, that support consciousness while the cerebellum appears to be experientially dead. How can we explain this? It turns out that, although the cerebellum has more neurons, its neurons are much less connected than those of the cerebrum. As a result, there is a lot more integrated information in the cerebrum than in the cerebellum, which is exactly what IIT predicts is important for consciousness.

IIT also explains why epileptic seizures and deep sleep do not involve consciousness, despite there being normal (or much higher than normal in the case of a seizure) levels of brain activity. The kind of brain activity we observe during seizures and deep sleep consists of "slow waves," highly regular series of bursts and silences which involve little integrated information. Again, this fits with the predictions of IIT.*

One of the most important practical aspects of consciousness research is to determine whether those in long-term comas who show no outward signs of consciousness are in fact having experience: whether they are "locked in," to use the medical jargon. IIT makes predictions about this: by scanning the brain to determine the level of integrated information, we can determine whether or not—according to IIT—the coma victim is conscious. Confirming these predictions is difficult, given that by definition coma victims are not very responsive. But insofar as it is possible, for example, by asking patients who have woken up, the predictions of IIT have stood up well.

At this stage, the evidence for IIT is far from conclusive. But suppose that one day IIT is as well confirmed as general relativity. Would this solve the "hard problem" of consciousness? In fact, it wouldn't. What a neuroscientific theory offers is *correlations* between certain physical states and states of consciousness. In the specific case of IIT, it is proposed that consciousness is correlated with integrated information: wherever you have

* As we discuss in chapter 4, some challenge the view that there is no consciousness in "dreamless sleep."

integrated information you have consciousness and vice versa. But IIT does nothing to explain why that correlation holds.

This leads to another way of understanding the distinction between the "easy problems" and the "hard problem" of consciousness:

- The Easy Problems: *Which* kinds of brain activity are correlated with consciousness?
- The Hard Problem: *Why* are certain kinds of brain activity correlated with consciousness?

IIT is one possible solution to the "easy" problems, but it leaves the hard problem unanswered.

IIT does not solve the hard problem on its own. But in conjunction with naturalistic dualism we do get at least a proposal for a solution to the hard problem. If a naturalistic dualist were persuaded of the truth of IIT, then her view would be that there is a fundamental psycho-physical law of nature—as fundamental as the law of gravity—that wherever you have integrated information you have consciousness.

The naturalistic dualist wouldn't claim to be able to explain *why* this fundamental psycho-physical law obtains, but, as we discussed above, that is simply the nature of fundamental laws. Newton wasn't able to explain why his laws obtain either. As fundamental principles of nature, the dualist's psycho-physical laws cannot be explained, but once we accept them we are able to predict the distribution of consciousness in the universe. As the philosopher Wittgenstein said, explanations come to an end somewhere. For some, explanation bottoms out with the laws of physics. For the naturalistic dualist, it bottoms out with the laws of physics *and* the psycho-physical laws. The psycho-physical laws sit alongside the basic laws of physics as simply the fundamental rules which govern our universe.

Can naturalistic dualism really be taken as a serious scientific proposal? Certainly naturalistic dualists have as yet provided nothing like the kind of detailed and empirically supported hypotheses we find in the physical sciences. But perhaps this is simply a reflection of how early we are in the scientific study of consciousness; perhaps we're still waiting for the "Newton of consciousness" to produce the simple equation that will capture the connection between body and mind. Given how hard it is to explain consciousness, some might argue that it's best to keep all options on the table.

In fact, most philosophers and scientists are not happy to keep dualism, naturalistic or otherwise, on the table. This is not because of any inherent mysteriousness in the *idea* of mind and matter interacting; there doesn't seem to be anything incoherent about the idea of there being basic psycho-physical laws of nature in addition to the basic laws of physics. Most philosophers reject dualism because they think that science has shown it to be false.

HAS SCIENCE PROVED THAT REALITY IS ENTIRELY PHYSICAL?

Religious texts are filled with stories of miraculous interventions in the workings of nature. When the Israelites were fleeing Egypt, God drew back the waters of the Red Sea so that they could walk along the seabed to safety. Some have argued that in certain freak circumstances such an event could have occurred naturally. Perhaps. But a more natural way of reading the story is that this was not something that would have occurred simply through natural events taking their course. Rather, it happened because God *interfered* in the workings of nature, making things happen that would not otherwise have occurred.

It is of course controversial whether or not God exists, and even if God exists it is controversial whether or not She intervenes in the world She has created. Some believers have a "deist" idea of God: a creator who lights the big bang but then sits back and lets things run their course. At the very least we can say that if God does intervene in the world, She doesn't act enough to make Her presence *obvious.*

But let's suppose we lived in a different world, in which God regularly intervened in the course of nature by miraculously healing diseases. What would such a world look like to medical scientists? We would expect there to be all sorts of changes in bodies which we could not explain through physical causes. Cancers would disappear, broken bones would mend, wounds would heal with inexplicable speed. Happenings that can't be explained through physical causes are known as "anomalous events." In the world we are imagining, God would make Her existence patently obvious to medical science through the presence of such anomalous events.

What I have just described seems to be a coherent world. But it isn't the world we live in. When medical scientists examine people in the real world, they don't seem to find such anomalous events, at least not very often. And the fact that we don't regularly find such anomalous events constitutes evidence that we don't live in a world in which there is a God who regularly intervenes in nature. This kind of evidence is never conclusive; it could be that doctors just keep missing the miracles for one reason or another. But as time goes on this seems more and more unlikely. The fact that ever more probing physiological investigation fails to reveal anomalous events gives us ever greater reason for thinking that there simply are no anomalous events. Either God doesn't exist or She keeps Herself to Herself.

What on earth does this have to do with dualism? The case many philosophers make against dualism is essentially the same form as the above case against an interventionist God. Imagine an immaterial mind were impacting on the brain every second of waking life, by initiating physical processes that caused limbs to move in accordance with the wishes of the conscious mind. When the mind wants to raise the right arm of the body, for example, it causes a change in the brain that will begin a causal process resulting in the raising of the arm. Each event that is directly caused by an immaterial mind will lack a physical cause. In this sense, each impact the mind has on the brain will be an anomalous event, a little miracle.

In other words, a nonphysical mind "intervening" in the brain is not much different in principle than a nonphysical God intervening in the body through healing. In both cases, something nonphysical—God or an immaterial mind—initiates a change in the physical world. In both cases, that change will have no physical explanation, and in that sense will be anomalous. Perhaps the only difference is that in the dualist case, the anomalous events would be much more frequent, due to the regular causal interactions between the mind and the brain. It would appear as though a poltergeist were playing with the brain.

It's hard to believe that such anomalous events would not show up in our neuroscience. There would be all kinds of things going on in the brain for which we had no neuroscientific explanation, precisely because they were caused by the interventions of the nonphysical mind. As we examined the brain, we would find "gaps" in the chains of physical causation where the mind had made a change in the brain. If a nonphysical God intervened in the world regularly, then Her presence would be obvious, because many things would happen which had no physical explanation. Similarly, if a nonphysical mind intervened regularly in the brain, then its presence would be obvious, because there would be a multitude of happenings in the brain each of which lacked a physical cause.

The problem for the dualist is that we don't seem to find anomalous events in the brain. Perhaps we wouldn't expect to when we knew very little about how the brain worked. However, although our neuroscientific understanding of the brain is very far from perfect, we now know how neurons work and we have a good understanding of the role different parts of the brain play in processing information and generating behavior. In none of this detailed investigation have we discovered anomalous events in the brain. Of course, it could be that we just keep missing them. But this seems more and more improbable as time goes on and they fail to turn up.

There are few things we can demonstrate with absolute certainty, and one never knows what will show up in future research. However, at this stage of inquiry, many philosophers and scientists feel reasonably confident that the failure to find anomalous happenings in the brain is strong evidence against dualism.

QUANTUM MECHANICS TO THE RESCUE?

How do naturalistic dualists respond to this worry? One speculative idea is that quantum mechanics might help us to locate the nexus between mind and body. In its essence, this idea has been discussed since the early days of quantum mechanics. In 1939, Fritz London and Edmond Bauer suggested that consciousness plays an essential role in quantum mechanics, and following on from this the Nobel Laureate Eugene Wigner wrote in 1961, "When the province of physical theory was extended to encompass microscopic phenomena, through the creation of quantum mechanics, the concept of consciousness came to the fore again: it was not possible to formulate the laws of quantum mechanics in a consistent way without reference to consciousness."[4] But it was not until the 1990s that the view was rigorously worked out by the American physicist Henry Stapp, who hoped that quantum mechanics might provide a way of reconciling human free will with physical causation.[5] More recently David Chalmers and his cowriter Kelvin McQueen have provided their own detailed interpretation of the idea.[6]

Quantum mechanics is basically a bit of mathematics which enables us to predict with great accuracy what's going to happen in the physical world. Due to the accuracy of its predictions, quantum mechanics is one of our best-confirmed scientific theories. Most of our modern technology, from computers to smart phones to GPS, is reliant on the predictive power of quantum mechanics.

The trouble is that nobody really knows what quantum mechanics is telling us about reality. We know it works: you run the equations and you can work out what's going to happen (or more precisely the objective probabilities of what will happen). But nobody knows what on earth is going on in physical reality to yield those results. This is not merely the point made above that science cannot explain why its most basic laws of nature hold. While Newton didn't explain why his theory of gravity was true, it was clear what the theory was telling us about the world. But when it comes to quantum mechanics, there is no consensus even on what the theory is proposing.

Why is quantum mechanics so hard to make sense of? There is not a single answer to this, rather there are a number of features of the theory that are hard to interpret. In chapter 4 we will discuss the phenomenon of "entanglement," whereby a pair of particles behave as a unified system even though they are separated by such enormous distances that it is not possible for a causal signal to pass between them. Stranger still, in quantum mechanics particles need not have precise locations and velocities, but can instead exist in "superpositions" of various locations/velocities. Nobody really knows what a superposition is, but we can think of it as a sort of refusal on the part of reality to be definitely one way or the other. A particle in a superposition between location X and location Y is in a strange state of being both at X and at Y while being definitely at neither.

This is all very odd. But still we haven't gotten to the feature of quantum mechanics that has most troubled philosophers and scientists. By far the strangest aspect of quantum mechanics is that *observation* seems to make a difference to how the universe behaves.

In its orthodox formulation of quantum mechanics, articulated by John von Neumann in 1932, there are two principles of nature governing at the subatomic level: the Schrödinger equation and the collapse postulate.[7] However, these laws of nature are inconsistent: they cannot both govern the same physical system at the same time. The Schrödinger equation is the one that implies all the weird stuff. It's perfectly happy for things to lack definite characteristics. As far as the Schrödinger equation is concerned, particles are neither *here* nor *there* but somehow both at the same time; radioactive substances neither *decay* nor *don't decay* but somehow manage to do both at the same time. The collapse postulate, in contrast, serves to eliminate such mysterious ambivalences. When it kicks in, reality has to decide. Particles are either definitely here or definitely there and never both; radioactive substances either have or haven't decayed.

What makes the difference between situations in which the Schrödinger equation rules and situations in which the collapse postulate runs the show? Here's the problem: on the face of it, it seems to depend on *what is being observed*. If a certain feature of reality is not being observed, the weird Schrödinger equation governs. But as soon as someone takes a peek, the collapse postulate takes over.

This is famously represented by the story of Schrödinger's cat, the thought experiment of the pioneer of quantum mechanics, Erwin Schrödinger (who of course the Schrödinger equation is named after). The poor cat is trapped in a box with a Geiger counter attached to a vial of poison and a small amount of radioactive substance. If the radioactive substance decays, the vial of poison will smash and the cat will die. If the radioactive substance doesn't decay, the cat will be saved. While the box is closed and the system unobserved, Schrödinger's equation rules the roost, with the result that the radioactive substance exists in a superposition of both decaying and not decaying, from which it follows that the cat is in a superposition of being both alive and dead. But as soon as the box is opened, the collapse postulate kicks in, ensuring that the superposition transforms into a definite value leaving the radioactive substance either decayed or not decayed and the cat either definitely alive or definitely dead.

It is certainly hard to make sense of a superposition. But it is even harder to make sense of the transformation of a superposition into a nonsuperposition being caused by observation. Why on earth would the fact that an observation is made change the world in this way? How does the universe know whether or not we're looking? Many scientists and philosophers don't like this at all and so try to find a way to avoid observation playing this crucial role in the theory. And this is where we start to get different interpretations of quantum mechanics.

The "Many Worlds" interpretation, formulated by Hugh Everett in 1957, has gripped the popular imagination.[8] According to this interpretation, different bits of the universe branch off into many possibilities, with each possibility genuinely existing in its own right. Why do physicists take this idea seriously? Because it removes the need for the collapse postulate. The collapse postulate is supposed to kick in when the weird superpositions of many possibilities transform into a single definite outcome. But on the Many Worlds theory, this never happens as all possibilities continue to exist in their own separate branches of reality. Talk of "superposition," on the Many Worlds theory, is just a way of describing the different possibilities existing at different branches: the cat is both alive and dead in different branches of the possibility tree. And because there is no change from superposed possibilities to a single definite outcome, there is no change from the governance of the Schrödinger equation to the governance of the collapse postulate, and hence no need for observation to play a role in making that transition. Problem solved! Except that the admission that everything that can happen will happen wreaks havoc on our commonsense judgments of probability.

Other interpretations of quantum mechanics maintain the transition from the Schrödinger equation to the collapse postulate but hypothesize more mundane mechanisms underlying the transition not dependent on the presence of observation. None of these interpretations has thus far managed to achieve consensus among physicists.

Is there another option? Is there a way of accepting what,

on the face of it, quantum mechanics seems to tell us: that observation makes a difference to physical reality? The problem is that "observation" doesn't seem to be a fundamental feature of reality, in which case it's hard to see how it could play such a significant role in our fundamental theory of the world. The fundamental things in physics are particles, fields, spacetime, and forces. How could "observers" be as fundamental as particles and fields?

But of course, there is a way in which we can make sense of observers being fundamental: by embracing dualism. According to dualism, conscious minds—or "observers" as we might call them—are basic features of the world, as fundamental as electrons or quarks. The basic idea of "quantum dualism"—as we might call it—is that it is the interaction of the nonphysical mind with the physical world that causes the change from the peculiar governance of the Schrödinger equation to the familiar governance of the collapse postulate. Schrödinger's cat exists in the indeterminate state of being both alive and dead only when there is no conscious mind interacting with the interior of the box via conscious observation. As soon as a human mind interacts with the relevant properties of the system, the superposition of being alive and dead is transformed into a definitely living or a definitely dead cat. If there are conscious minds interacting with the physical world, then ascribing a fundamental role to observation is perfectly intelligible.*

Let us think about this theory in a little more detail. The first

* What about the consciousness of the cat itself? While Descartes believed that animals were unfeeling mechanisms, this is not a view shared by any naturalistic dualist I've ever come across. To get around this problem, we could make the cat unconscious, or indeed take the cat out altogether and just focus on the transformation of the radioactive substance from superposition to definite state.

point of contact between the mind and the physical world is in the brain. According to quantum dualism, the conscious mind impacts on the brain by transforming certain brain events from a state of superposition to possessing a definite characteristic or "property." It could be, for example, that before the mind interacts with it, a certain part of my brain exists in a superposition between the following two physical possibilities:

- *Physical property X:* A state of affairs that would cause my arm to raise.
- *Physical property Y:* A state of affairs that would not cause my arm to raise.

By interacting with the relevant part of the brain, the mind causes the superposition to resolve into either property X or property Y, with the result that my arm either goes up or stays down. In virtue of resolving the superposition, the conscious mind plays a role in the dynamic process that causes my arm to move or not to move.*

The attraction to the dualist of this interpretation of quantum mechanics is that it provides a causal role for the conscious mind at the heart of physics. In the last section, we discussed the worry that the actions of a nonphysical mind on the brain would show up as anomalous brain events lacking physical explanation, and that neuroscience seems to show no sign of such anomalous events (which opponents of dualism say gives

* I am simplifying a little. On Chalmers and McQueen's initial form of the view, properties of the brain that are correlated with consciousness do not themselves enter into a superposition, although brain properties existing a moment earlier do. On later modifications of the view, even the brain properties that are correlated with consciousness briefly enter into superpositions.

us grounds for doubting the existence of nonphysical minds). But if quantum dualism is true, the immaterial mind plays a role complementary to, rather than at odds with, physics. On the face of it, physics suggests that something in the vicinity of observation changes superpositions into definite values. The quantum dualist merely adopts this role for the conscious mind. As David Chalmers once remarked to me, "If you wanted a scientific theory that makes room for the conscious mind to play a fundamental role, you couldn't hope for something better than quantum mechanics."

Quantum dualism is a fascinating approach that deserves to be explored in more detail. A kind of cultural prejudice has prevented it from being taken more seriously and consequently it is not as well developed as other interpretations of quantum mechanics. Indeed, as Chalmers and McQueen are keen to point out, there is a kind of vicious circularity in the way that dualism and the dualist interpretation of quantum mechanics are standardly rejected. Philosophers typically reject dualism because of the interaction problem: the difficulty of making sense of how the conscious mind impacts on the physical world. But if quantum dualism is true, this problem is potentially avoided, as consciousness plays a role in the fundamental dynamics of the physical world. Meanwhile, scientists reject quantum dualism on the grounds that it involves an allegedly discredited theory, namely dualism. And hence it very much looks like dualism is being rejected by assuming the falsity of quantum dualism, while quantum dualism is being rejected by assuming the falsity of dualism. The situation is rather like the religious believer who believes God exists because it says so in the Bible, while believing the Bible because it's the word of God. In both cases, the argument goes around in a circle, establishing nothing but the prejudices of its proponents.

Having said that, as things stand, I don't think that quantum dualism, at least as it has been developed by Chalmers and McQueen, provides *enough* of a causal role for the conscious mind. Above I described how, according to quantum dualism, consciousness resolves the superposition between properties X and Y, causing my arm to either go up or remain down. You might have imagined that the conscious mind *freely decides* which property—either X or Y—the superposition is going to resolve into, and thereby decides whether the arm will go up or down. However, to suppose this would lead to a tension with the physics. This is because in standard formulations of quantum mechanics there is a purely physical principle—the Born rule—that determines the probabilities of which definite properties emerge from a superposition. If we want to say that consciousness determines which brain properties emerge from a superposition, then the quantum dualist would have to find some way of ensuring that the free actions of conscious minds yield the probabilities implied by the Born rule. Perhaps this could be done, but it's not something that any quantum dualist has thus far worked out how to do.

Instead, Chalmers and McQueen propose a very minimal role for consciousness. The interaction of the conscious mind with the brain causes the brain to stop existing in a given superposition, but after that the Born rule takes over to determine the probabilities of which definite properties emerge from the superposition. The role of the mind is merely to say, "Let superpositions be resolved!" and then physics in conjunction with random chance determines what actually occurs. To return to our example, my conscious mind makes it the case that my arm will subsequently be either definitely up or definitely down, but it is physics and random chance that determine which of those two possibilities will become actual.

In other words, according to quantum dualism as formulated by Chalmers and McQueen, consciousness does not determine what actions we do, or what words we say, or in general what kind of impact a person has on the world. This is far less than the role we intuitively suppose the mind to have. Of course, perhaps our intuitions about what consciousness does are wrong. Still, an important aim of quantum dualism is to preserve the commonsense idea that my mind is the cause of many of my actions. And it's not clear that the theory of Chalmers and McQueen really lives up to this goal.*

THE VALUE OF SIMPLICITY

I don't think we should stop thinking about dualism and trying to work out if it's consistent with our scientific picture of the world. Despite the contempt with which it is often held in contemporary culture, it is far from clear that science has shown dualism to be absurd. Nonetheless, it should be treated as the last resort. Even if it ends up not being in tension with science, we should be wary of dualism on the grounds that it's significantly less simple than other theories of consciousness.

One of the most important principles in science and philosophy is "Ockham's razor," named after medieval philosopher William of Ockham (1287–1347), who wielded it with ferocity against the extravagant theories of his opponents. Ockham's razor is quite simply the principle that, all things being equal,

* There are other problems with the view, most notoriously the "quantum Zeno effect," discussed in Chalmers and McQueen's article "Consciousness and the Collapse of the Wave Function."

we should try to make our theories of reality as simple as possible. Einstein put it as follows:

> It can scarcely be denied that the supreme goal of all theory is to make the irreducible basic elements as simple and as few as possible without having to surrender the adequate representation of a single datum of experience.[9]

In other words, if theory X and theory Y can both explain all the data, but theory X is simpler or more parsimonious (i.e., postulates fewer entities) than theory Y, then theory X is the one we should go for. For example, if theory X postulates twelve kinds of fundamental particle and theory Y postulates thirteen kinds of fundamental particle, then theory X is the one we ought to go for (assuming there is no predictive advantage secured by the additional postulation).

It is somewhat mysterious *why* simplicity is to be preferred. Why on earth should simpler theories be more likely to be true than complex ones? And yet without this principle, scientific inquiry is impossible. This is because, whenever we have some data, there will always be an infinite number of hypotheses that can account for that data. Don't believe me? Consider the standard model of particle physics: an impressive theory able to account for a huge range of observable data. Now consider another theory, call it "the angelic model," which postulates everything that the standard model postulates but also postulates an impotent angel, that is to say, an angel that observes the universe without being able to do anything. The standard model and the angelic model make exactly the same predictions; given that the angel can't do anything, its existence would make no difference to what we observe. Now consider the "dual angelic model," which adds another impotent angel to the postulations, or the "triple angelic model," which postulates three impotent angels, and so on ad infinitum. What we end up with through this procedure is an infinite number of theories that can't be distinguished through observation.

You might be thinking that this discussion is a little silly. Of course the basic standard model is the one we ought to go for, as these extra angels add nothing to the theory. I agree. But this reasoning in support of the standard model over the angelic alternatives implicitly appeals to Ockham's razor: the standard model is to be preferred not because it is able to account for more observations than the angelic alternatives—all of these theories are on a par in terms of their predictive power—but because it's simpler.

Most of the time choosing the simplest theory is pretty straightforward, and so we don't really notice that we're using Ockham's razor. But there are cases in the history of science when the principle has played a clear and decisive role in theory choice. A nice example comes from Einstein's special theory of relativity. Einstein's theory had the same predictive power as the view of Hendrik Lorentz that preceded it. Both theories were able to account for the fact that all nonaccelerating observers, no matter how fast they are moving, will measure the speed of light to be the same. This was the startling finding of Albert A. Michelson and Edward W. Morley, discovered in their famous experiment of 1887, and both Lorentz's theory and Einstein's theory have a way of explaining it. However, on Lorentz's view, this observation was not what it seemed. The speed of light only *appeared* to be the same in all frames of reference. Lorentz postulated forces operating on our tools of measurement—clocks and measuring rods—that gave the *illusion* that the speed of light remained the same relative to different observers, when in reality it is changing.

Einstein's theory, in contrast, gave a much simpler account of the data, which dispensed with the need for such forces. According to the special theory of relativity, the speed of light appears to be the same for all nonaccelerating observers . . . because it is! The fact that the physics community almost universally accepts Einstein over Lorentz, despite the fact that their theories are observationally equivalent, shows that Ockham's razor is very much at work in the scientific method.

Returning to the question at hand, if dualism can't account for mind-brain interaction, then it should be rejected for that reason. But even if it can, dualism is less simple and hence less attractive than other solutions to the problem of consciousness. In the next chapter, we will explore *materialist* views, according to which consciousness can be accounted for in terms of the electrochemical process of the brain. If this view can be made to work, then it offers hope of a more parsimonious explanation of consciousness. If consciousness can be explained in terms of the brain, then the postulation of an immaterial mind is surplus to requirements.

We shouldn't believe in immaterial minds unless we really have to. The rest of this book will examine the prospects for nondualistic theories of consciousness, starting in the next chapter with materialism.

3

Can Physical Science Explain Consciousness?

Let's return to Susan, your best friend whose brain we were contemplating at the start of the last chapter. Staring down at the trillions of neural connections we again ask the question: Where is *Susan*? The topic of the last chapter was dualism. If dualism is true, Susan is an invisible immaterial mind, and the body and brain you are now looking down on is merely the tool Susan uses for interacting with the world. The topic of this chapter is *materialism*, and materialists believe that when it comes to Susan what you see is what you get. There are no immaterial or invisible parts to Susan; Susan just is the functioning body and brain you are currently beholding.

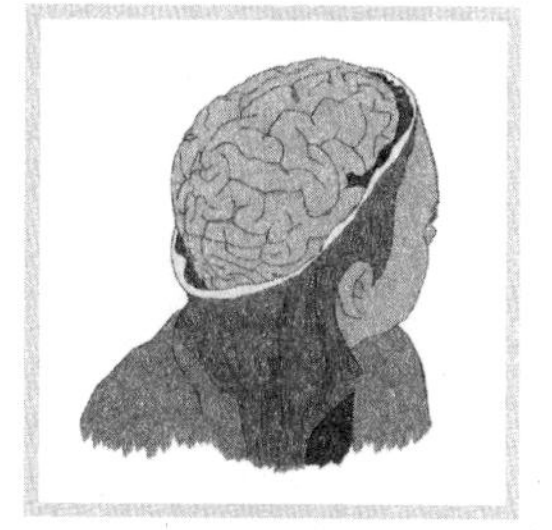

What about Susan's feelings and experiences, her pleasures and pains, her sensations of colors and sounds and smells? For the materialist, the inner subjective world of experience is to be explained in terms of the chemistry of the brain, in something like the way the wetness of water is explained in terms of its molecular structure. It is generally agreed that, in contrast to the case of water, we don't yet have a satisfying explanation of how this happens. But materialists have faith that one day scientists will crack the mystery and bring consciousness fully into the scientific story of the world.

Materialism is an attractive view, dispensing as it does with the sense of magic and mystery that the topic of consciousness tends to invite. So let's examine it more closely.

WHAT'S THE POINT OF PHILOSOPHY?

The title of this chapter is "Can physical science explain consciousness?" Can philosophy really answer this question? There will be many who are skeptical of "ex cathedra" pronouncements of philosophers such as myself concerning what physical science can or cannot do. Indeed, there is in our culture at the moment a fair amount of skepticism—or perhaps a better word is "bemusement"—at the idea that philosophers might have anything to contribute to the scientific project of uncovering the nature of reality. Such skepticism is presumably rooted in the fact that, in general, philosophers reach their conclusions without actually performing any experiments or carrying out any observations. In contrast to the natural scientist, the principal activity of the philosopher is *thought*. Engaging in philosophy does not ordinarily require one to leave one's armchair, or even to get out of bed! (I myself am proud to follow in the great tradition of Descartes, Keynes, and Churchill of working in bed. . . .)

Here is how the cosmologist Lawrence Krauss expresses this skepticism:

> As far as the physical universe is concerned, mathematics and experiments, the tools of theoretical and experi-

> mental physics appear to be the only effective ways to address questions of principle. . . . To those who wish to impose their definition of reality abstractly, independent of emerging empirical knowledge and the changing questions that go with it, and call that either philosophy or theology, I would say this: Please go on talking to each other, and let the rest of us get on with the goal of learning more about nature.[1]

This kind of worry is entirely understandable. It is quite intuitive to think that in order to find out about the world, you have to go and look at it, by making observations or conducting experiments. If we had a magical sixth sense, then perhaps by entering into a deep meditative trance we could plumb the depths of reality from the comfort of an armchair. Sadly, we don't.

Indeed, these kinds of worries about the potential of philosophy to contribute to finding out about reality are not confined to science. In their polemical book *Every Thing Must Go,* philosophers James Ladyman and Don Ross rail against the "pseudo-scientific metaphysics" of their fellow philosophers:

> Suppose that the Big Bang is a singular boundary across which no information can be recovered from the other side. Then, if someone were to say that "The Big Bang was caused by Elvis," this would count . . . as a pointless speculation. There is no evidence against it—but only for the trivial reason that no evidence could bear on it at all. [2]

This book concerns the science of consciousness rather than the origins of the universe. But Ladyman and Ross intend their point to be entirely general, and, in the context of discussing consciousness, they argue that only the physical sciences of cognitive science and neuroscience can shed light on the nature of the mind. And, of course, Patricia Churchland, whom we came across in the first chapter, has little time for her fellow philosophers who theorize about consciousness independently of the scientific study of the brain. In her terms, such philosophers are lazy "nay-sayers," holding back the onward march of scientific progress. She describes their methods thus:

> No equipment had to be designed and maintained, no animals trained and observed, no steaming jungle or frozen tundra braved. The great advantage of nay-saying is that it leaves lots of time for golf.[3]

Despite all this bluster, it is self-evident that we can know at least some things about reality just by thinking. Here's an example. I can know without leaving my armchair that nowhere in the universe are there any square circles. How do I know this? Because the very idea of a square circle involves contradiction, and anything that involves contradiction is impossible. Similarly, I don't need to perform an experiment to know that, no matter how weird and wacky the universe turns out to be, it doesn't contain planets which aren't planets, or rocks that are both blue all over and not blue at all.

How is it possible to draw such general conclusions about the universe without even standing up? By relying on the following most fundamental law of science in its most general sense:

> *The Law of Noncontradiction*—Any hypothesis which involves a contradiction is false.

This is the principle which mathematicians and logicians rely on in order to reach their conclusions. One proves a theorem in mathematics by demonstrating that its denial involves contradiction. And the philosopher can make use of exactly this method: if it can be shown that a rival philosophical hypothesis involves contradiction, then that hypothesis can be ruled out.

Of course, the examples of contradictory theories of reality given above are all completely obvious and uncontroversial. If the only thing philosophers were able to demonstrate is that the universe doesn't contain square circles or that nothing is both blue and not blue, then their method would be quite impotent. However, some theories are contradictory in a way that is subtle and far from obvious, and uncovering these more subtle contradictions may require the distinctive skills of the armchair philosopher.

Consider the case of time travel. In the classic 1980s sci-fi film trilogy *Back to the Future,* we find many instances of the characters *changing the past.* In the first movie, our hero Marty McFly nearly stops his parents meeting, and consequently finds his hand starting to disappear as the past is "rewritten." In the sequel, the obnoxious neighbor Biff steals the time-traveling DeLorean and visits his younger self to deliver a sports almanac recording results for the next sixty years, thus creating an alternative history in which Biff is a millionaire.

On the face of it, such plotlines seem perfectly intelligible. We know we're not watching something *real,* but what we're watching seems to be at least *possible.* A film about square circles—if we can even imagine such a thing—would be unwatchable, but films about changing the past seem to make perfect sense. In spite of this, there is a broad consensus among philosophers that such cases of "changing the past" are simply incoherent. To see this, we need first to learn a little philosophy of time.

KILLING GRANDDAD

There are two theories that dominate the philosophy of time: presentism and eternalism. Presentism is the theory of common sense. According to it, only the present moment exists: events in the past—such as the Norman conquest of England in 1066—have ceased to be, while events in the future—such as the first human landing on Mars—are yet to be. For the presentist, there is something very special about this second of this day, as events that happen at this time are the only ones that really exist.

Eternalism maintains, in contrast, that all events in time are equally real. Imagine you were to look down at reality from a God's eye point of view, with all events in time laid out in order, from the big bang to final "heat death" when all the energy of the universe has been used up. Eternalism tells us that from this perspective, there would be no special moment in time marked "now," just as there is no special location marked "here." Indeed, the word "now," according to eternalism, functions like the word "here": just as "here" is the name for where we are in space, so "now" is just the name for when we happen to be in time. And therefore, for a hypothetical creature outside of space and time, "now" and "here" would have no meaning.

To put it another way, the eternalist defends a kind of *temporal egalitarianism*. If you think there's something special about the present moment then you're guilty of *chronological chauvinism*, or "time racism": privileging without justification your own time over that of others. The thoughts and feelings of the Norman warriors of 1066 are no less real than your thoughts and feelings; the bodies of the Martian colonists no less solid to touch. Due to our position in time we are unable to see past and future events, but that doesn't make them unreal; it just makes them somewhere—or rather some*when*—else.

Now that we have a bit of philosophy of time under our belt, we can return to the topic of time travel. The first thing to note is that if presentism is true, time travel isn't even possible. According to presentism, the past and future don't even exist; and you can't go somewhere that doesn't exist. I can't go to the planet Vulcan. Why not? Because Vulcan doesn't exist. Similarly, if presentism is true, I can't go to 1066. Why not? Because 1066 doesn't exist (although, unlike Vulcan, it *used to* exist). If presentism were true, then getting in your time machine and setting the controls for the past would be a suicide mission: you'd be heading straight into nonexistence.

So we need the eternalist theory of time to be true for time travel to be possible. The past and present need to actually be there in order for us to be able to visit them. That's okay, as many physicists believe that eternalism, although arguably in tension with common sense, is strongly supported by Einstein's theory of special relativity (Einstein's theory fits better with eternalism because according to relativity there is no special, privileged "now"). And, indeed, within the context of eternalism, time travel is perfectly coherent. Again adopting our "God's eye perspective" for the sake of illustration, we can imagine looking down at all the events in time laid out in order, and notice that one causal sequence—involving a DeLorean approaching 88 miles per hour—begins in the year 1985 and then continues in the year 1955.

Eternalism allows for the possibility of time travel; but what it does not allow for is the possibility of time travel that involves *changing the past.* Don't all time travel stories involve changing the past? Not necessarily. Consider, for example, the original *Terminator* movie, or the wonderfully complicated Spanish movie *Timecrimes.* These movies tell stories in which people from the future visit the past but, unlike *Back to the Future,* they don't involve visitors from the future *changing* the past. In *The Terminator,* the cyborg played by Arnold Schwarzenegger travels from the year 2029 to the year 1984 in order to kill Sarah Connor, who will one day give birth to the savior of the humans fighting against the evil machines. If the Terminator had succeeded in its ambitions, then we would have had a film in which there are "two" versions of history:

> In the "first" version of history, Sarah Connor gives birth in 1985.
>
> In the "second" version of history, Sarah Connor dies in 1984 and so never gives birth.

It is when you get two versions of history—the second one created by the interventions of time travelers—that you have a case of changing the past. This is what we find in *Back to the Future:*

> In the "first" version of history, Biff ends up having a modest income.
>
> In the "second" version of history, Biff ends up being a millionaire.

But this is in fact not what we find in the actual *Terminator* movie. The Terminator fails to kill Sarah Connor, and because of this the audience is not shown multiple versions of history.

The consensus of philosophers is that time travel films involving multiple versions of history are incoherent. In describing such films, we talk, as I have done above, about the "first" version of history and the "second" version (maybe more if there are lots of interventions . . .). But when we are considering the eternal facts of time, talk of "first" and "second" versions of history makes no sense. Returning to the God's eye perspective, when you look down at all the events in time, either the pyramids exist or they don't, either Biff is a millionaire or he isn't. Talk of "change" makes no sense when you are looking down at all of time.

Here's a way of making the point without the theological metaphor. The crucial question to ask is: *When* did Biff change the past, replacing one version of history with another? Was it in 1955? Well, no, because in the "first" version of history—whatever that means—Biff wasn't visited by his future self in 1955. Was it some other time than 1955? No, because in the "second" version of history, that's precisely the time at which the past changes. What we want to say is that in the "first" 1955 old Biff did not visit young Biff and in the "second" 1955 he did. But there is no way to cash out what "first" and "second" mean here. Careful thought—what philosophers are good at—reveals that the very idea of "changing the past" is nonsense.*

* Some philosophers have entertained the possibility of a second dimension of time—"hypertime"—in order to accommodate multiple versions of history, so that we would literally have multiple 1955s stretched out along the hypertime dimension. This is indeed a coherent possibility, although we have no reason to think there is such a thing as hypertime and the makers of time travel films don't seem to commit to such a thing. To be more precise, the claim for which there is consensus is: changing the past is incoherent assuming a single time dimension.

THE GREATEST PHILOSOPHER WHO EVER LIVED

Time travel is the stuff of science fiction. But there are also cases in the history of science in which armchair reflections have yielded serious conclusions about the nature of reality. As we discussed in chapter 1, prior to the scientific revolution, theorizing about the universe was dominated by the ideas of the ancient Greek philosopher Aristotle. The scientific revolution swept away these ideas; the observations of Copernicus proved, for example, that the earth was not, as Aristotle had supposed, in the center of the universe. What is often overlooked, as we began to discuss in chapter 1, is that Galileo refuted one of the central claims of Aristotle's physics not with observation or experiment but with a philosophical thought experiment.[4]

The particular bit of Aristotle's physics in question was one of the most intuitive: heavy objects fall faster than lighter ones. More specifically Aristotle believed that the speed at which a falling object accelerates to the ground is proportional to its mass. For example, a bowling ball that weighs seven kilos will fall fourteen times faster than a football that weighs half a kilogram.

It is natural to assume that heavier things fall faster than lighter things, and indeed following Aristotle people believed it for thousands of years. However, it's simply not true. As Galileo proved, putting aside factors such as air resistance, all falling objects—no matter how heavy or how light—accelerate to the ground at exactly the same rate. If we could somehow get rid of air resistance, and we dropped a Ping-Pong ball, a football, and an elephant from a great height at the same time, all three would hit the ground at precisely the same time.

Legend has it that Galileo proved this by dropping weights from the Leaning Tower of Pisa. But historians doubt that this experiment ever happened. And indeed if it had occurred, then the objects would not have hit the ground at the same time due to the effects of air resistance. A dramatic experimental confirmation of Galileo's view took place during the Apollo 15 mission to the moon of 1971. At the end of the mission Commander David Scott dropped a feather and a hammer to the surface of the moon. Lo and behold, in the absence of air resistance, both objects hit the lunar surface at precisely the same time. Galileo had been vindicated by experiment.

However, such experimental proof was entirely unnecessary, because Galileo had already decisively demonstrated his position to be correct with a thought experiment. From the comfort of his armchair Galileo proved that Aristotle's view, the view which everyone had believed for thousands of years, involved contradiction and hence could not possibly be true. If you're sitting comfortably in your armchair, we can perform Galileo's thought experiment together.

In his thought experiment, Galileo asks us to assume that Aristotle is correct: heavier objects fall to the ground faster than lighter objects. Now imagine we drop two objects from a great height; for the sake of vividness let's stick with the elephant and the bowling ball. Galileo added another crucial detail to the thought experiment: before we drop the elephant and the bowling ball, we chain them together.

Now ask the following question: Would the elephant fall faster in this situation than it would have done if it weren't chained to the bowling ball? In other words, does the fact that the elephant is chained to the bowling ball slow the elephant down or speed it up?

The genius of Galileo was that he realized, through pure

rational reflection, that—assuming Aristotle's physics—this question has two contradictory answers.

Answer A: The fact that the elephant is tied to the bowling ball slows it down

Given that the bowling ball is much lighter than the elephant and hence falls much more slowly than the elephant, the chain between the two will become taut. This will have the result of slowing the descent of the elephant down slightly, and hence the elephant will fall slightly slower than it would have done if it were not chained to the bowling ball. In this way, the bowling ball acts as a kind of parachute for the elephant.

Answer B: The fact that the elephant is tied to the bowling ball speeds it up

Given that the elephant and the bowling ball are joined together, we may consider them as one object, made up of an elephant, a bowling ball, and a chain joining them. The weight of that object will be the combined weight of the elephant, the bowling ball, and the chain. So obviously the weight of that combined object will be greater than the weight of the elephant on its own. It follows that the combined object (elephant + bowling ball+ chain), which includes the elephant as a part, will fall faster than would the elephant on its own.

Both answer A and answer B follow from Aristotle's commonsense assumption that heavier objects fall faster. And yet answer A and answer B are contradictory, and hence—by the law of noncontradiction—they can't both be true. So long as we suppose the weight of an object affects the speed of its descent, we are going to run up against this kind of contradiction. The only way to remove the contradiction is to suppose that all objects—whatever their weight—fall to earth at precisely the same speed. And we can know this to be true, as Galileo knew it to be true, by pure reason.

Galileo lived in an age when science and philosophy were not clearly distinguished. And it was as a philosopher, using philosophical methods, that he refuted this crucial aspect of Aristotle's physics. There is a deep irony here. It is sometimes claimed that the scientific revolution, and the great progress which followed it, have rendered philosophy impotent as a tool for understanding the natural world. And yet the father of the scientific revolution is in fact the great vindicator of the philosophical method. Galileo is one of the few philosophers to have produced a philosophical argument which nobody has ever disputed; and with this argument he transformed our understanding of the physical world.

WHAT IS IT LIKE TO BE A BAT?

We have had a long digression via square circles, time travel, and Aristotelian physics. The purpose of this digression was to persuade the reader that philosophy has a role to play in the project of finding out what reality is like. If the philosopher can show that a certain hypothesis about reality contains a subtle contradiction—as can be shown in the hypothesis that the past can be changed, or that heavier objects fall faster—then she can demonstrate that the hypothesis in question cannot possibly be true. This is precisely the methodology I will use against materialism. My central claim is that materialism can't possibly be true for the same reason that there can't be square circles or time-traveling patricide: the materialist theory of consciousness involves a contradiction.

Here are two things we know about consciousness:

- Consciousness involves *qualities:* the redness of a red experience, the feel of an itch, the rich taste of biting into chocolate.
- Consciousness is *subjective,* in the sense that knowledge of a given state of consciousness involves *adopting the perspective* of someone who has that conscious state.

We spent a lot of chapter 1 focusing on the first of these characteristics of consciousness. Let us now dwell a little on the second.

Physics aspires to describe the world in entirely *objective* terms, in terms that could be understood by anyone, whatever his or her life experience. Visiting aliens might have incredibly different sensory organs, and as a result they might not be able to understand or appreciate our art or our music. But if they are intelligent enough to understand mathematics, then they will be able to grasp our physics. Physics aspires to what the philosopher Thomas Nagel called "The View from Nowhere."[5]

Consciousness cannot be understood from The View from Nowhere. To be conscious is to adopt a specific perspective, and hence the conscious life of a particular organism can only be understood by adopting the perspective of that organism. In his groundbreaking paper "What Is It Like to Be a Bat?," Nagel argued that no matter how much we learn about the biology or neurophysiology of a bat, we will never understand its consciousness. This essential limitation arises from the fact that we cannot adopt the perspective of a creature that echolocates its way around its environment. There will always be something we cannot grasp about bats, namely, *what it is like to be a bat.*

In the 1990s cult movie *Being John Malkovich,* the characters find a tiny door in an abandoned office building behind which is a small tunnel. As the characters squeeze into the tunnel and clamber along it a little way, they find themselves starting to accelerate . . . faster and faster and faster . . . until suddenly . . . they find themselves experiencing *what it is like to be* the real-life actor John Malkovich: looking out of his eyes and hearing out of his ears. In other words, the tunnel magically allows the characters to adopt the conscious perspective of John Malkovich. We can maybe imagine a similar film called *Being a Dog,* where the characters clambering down the tunnel find out what it's like to be a dog, although even our ability to take up a dog's perspective is limited (What must it be like to smell your way around the world?). But no director would know how to make an analogous film called *Being a Bat.* The radical incommensurability of human and bat perspectives means that we have no idea what it's like to be a bat.

What has all this got to do with the coherence of materialism? After all, the reason we are unable to adopt a bat's perspective is that human beings have a very different *physical nature* compared to bats. Perhaps in the future it will be possible for a blind person to have an operation that installs sophisticated sonar into her body, allowing her to echolocate her way around the world. Such a person might, through this alteration to her physical nature, be enabled to discover what it's like to be a bat. Doesn't this all suggest that consciousness is to be explained in terms of the physical facts of biology?

Fair point. All I've been doing so far is trying to characterize the two essential characteristics of consciousness: the *qualita-*

tive (consciousness involves qualities) and the *subjective* (you can only understand my consciousness if you can adopt my perspective). The problem for materialism is that its conceptual resources are not up to the task of characterizing these features of consciousness. Physical science tries to give a purely *objective* characterization of reality, a characterization that can be grasped by anybody regardless of his or her perspective. To say that reality can be exhaustively described in such terms is to say that there are no subjective properties, i.e., no features of reality that can be grasped only from a certain perspective. Conversely, to say there are subjective properties—properties which, by definition, can be grasped only from a certain perspective—is to say that reality cannot be exhaustively described in purely objective terms. Materialists who claim *both* that reality can be exhaustively described in the objective vocabulary of physical science *and* that there are subjective properties are quite simply contradicting themselves.

Likewise, physical science tries to give a purely *quantitative* characterization of reality, a description involving only mathematical terms. Mathematical concepts—such as the concept of "two" or the division function—are radically different to qualitative concepts—such as "yellow" or "sour"—and the latter cannot be defined in terms of the former. To say that reality can be exhaustively described in purely *quantitative* terms is to say that there are no *qualitative* properties. Conversely, to say that there are qualitative properties is to say there are features of reality that cannot be captured in purely mathematical terms. Materialists who claim both that reality can be exhaustively captured in the quantitative language of physical science and that there is quality-rich consciousness contradict themselves.

Galileo used a thought experiment to demonstrate the incoherence of Aristotelian physics. Contemporary philosophers have devised a variety of thought experiments to demonstrate the incoherence of materialism. In the next section we will discuss the most famous: that of Black and White Mary.

THE TALE OF BLACK AND WHITE MARY

Black and White Mary has become one of the most famous nonexistent characters in Western philosophy. The writer David Lodge devoted a whole chapter of his novel *Thinks . . .* to riffing on the story of Mary. And philosophers certainly can't get her out of their heads. So great has been the obsession with Mary that one volume of essays devoted to the thought experiment is titled *There's Something About Mary*.[6]

The thought experiment we are about to discuss is used in a philosophical argument known as "the knowledge argument." The aim of the knowledge argument is to demonstrate that the knowledge of consciousness provided by the physical sciences is *necessarily incomplete*, that it will always leave something out. This style of argument is not new. A kind of knowledge argument was proposed in the seventeenth century by Gottfried Wilhelm Leibniz (1646–1716), a genius mathematician and philosopher who was hated by Newton for discovering calculus at the same time Newton himself did.

Leibniz pressed his version of the knowledge argument with his own intriguing thought experiment. Leibniz imagined we could somehow increase the size of the brain many times, so that it was big enough for us to climb inside "as one enters a mill." No matter how much we wandered around the giant brain, examining its workings, we would find only "pieces which push one against another, but never anything by which to explain a perception [by which Leibniz meant a conscious experience]."[7] Knowledge of mechanism, Leibniz concluded, can never yield knowledge of the conscious mind.

Leibniz's thought experiment is suggestive but far from convincing. It is a vivid way of pressing the intuition that mere mechanical workings could never add up to consciousness, but it doesn't really provide an *argument* for this claim. Moreover, if there were a neuroscientific explanation of consciousness, it is likely to be in terms of holistic features of the brain, such as its connectivity or its predictive processing, which one might not be able to take in simply by examining little bits of the brain in isolation. What we need is a thought experiment to show that whatever information we glean from reading neuroscience, it alone will never be enough to explain consciousness.

This is what we find in the most famous contemporary version of the knowledge argument, due to the Australian philosopher Frank Jackson. Jackson came up with his argument in an article in 1982, and there have subsequently been countless articles and books discussing it.[8] Materialists are certain that *something* is wrong with it, but there is no consensus on exactly what. Ironically, despite having formulated the most well-known argument against materialism, Jackson later lost faith in his argument and for the past twenty years has been a committed materialist.

Let's begin with the thought experiment itself, and then we'll get on to what it's supposed to show.

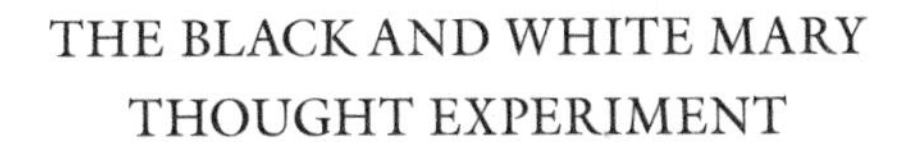

THE BLACK AND WHITE MARY THOUGHT EXPERIMENT

Mary is a genius brain scientist. However, for some reason that is never quite made clear, she has spent all of her life in a black-and-white room, and as a result has never seen any colors beyond black and white and shades of gray.

Despite this limitation, Mary is able, through watching lectures on her black-and-white television, to learn neuroscience. And learn she does. Mary comes to know *all the physical facts* involved in color experience. Mary knows precisely what happens in the body and brain when, for example, someone sees a red tomato. She knows the wavelength of light that enters the retina, the changes that makes to the sensory cells in the eye (known as cones and rods), how this information is transferred to various regions of the brain, and how it ultimately causes various behavioral responses, e.g., reaching for the tomato. In other words, Mary learns everything neuroscience could ever teach us about color experience without ever actually seeing any colors herself (apart from black and white and shades of gray).

The story has a happy ending. One day Mary is liberated from her black-and-white room and, for the first time in her life, experiences a world of color. We might imagine her, as she steps out into the world, encountering a lemon tree adorned with bright yellow lemons on the threshold of her black-and-white room. According to Jackson (or rather, according to Jackson in the days when he accepted his own argument), Mary at this point learns something new: she learns *what it's like to have a yellow experience.*

The End.

What is this peculiar fable supposed to show? Why would anyone think it demonstrates the falsity of materialism? Despite the catchy story, the point of the argument is a little subtle and often misunderstood. I have sometimes come across philosophers who think the argument is supposed to be understood as follows:

1. If materialism is true, then anyone who knows enough neuroscience will be able to experience colors.
2. Mary knows all the relevant neuroscience but cannot experience colors.
3. Therefore, materialism is false.

This is clearly a bad argument. Nobody would think that learning about the physics of black holes will create a black hole inside you. So why should we suppose that learning about the neuroscience of color experiences will create color experiences inside you? In fact, this is not how the argument is supposed to be read.

The focus of the argument, as its name suggests, is *knowledge*. To continue with our analogy, imagine we already have the complete and final theory of black holes. While we wouldn't expect that learning this theory will turn you into a black hole, we would expect mastery of it to give you complete knowledge of the nature of black holes.

Holding this example in your mind, let's return to the Mary story. According to the story, Mary in her black-and-white room knows everything physical science can possibly tell us about color experiences. If materialism is true and neuroscience is able to give us a complete theory of the nature of color experience, then what pre-liberation Mary has learned is the complete and final theory of color experience. It follows that she ought to know all of the essential features of color experience, just as someone who knows the complete and final theory of black holes ought to know all the essential features of black holes. And yet, when she leaves her black-and-white room, Mary learns about some new essential features of color experiences: she learns what it's like to have color experiences. It follows that a neuroscientific theory of color experience is *necessarily incomplete.*

Here's the argument:

The Knowledge Argument

1. If materialism is true, then Mary in her black-and-white room has a complete and final theory of color experience.
2. If Mary in her black-and-white room has a complete and final theory of color experience, then it

shouldn't be possible for her to learn about some new essential features of color experiences.

3. And yet, when Mary leaves her room, she does come to learn about new essential features of color experiences: she learns about what it's like to have color experiences.
4. Therefore, Mary in her black-and-white room can't have had a complete and final theory of color experiences and materialism is false.

If neuroscience is unable to give us a complete theory of color experience, what does it leave out? It leaves out the *subjective qualities* involved in color experience, those qualities we are directly aware of when we see colors. When we have a yellow experience, we are directly aware of the yellowish quality that defines it. If a neuroscientific description of the brain could fully characterize this quality, then Mary in her black-and-white room could come to know what it's like to have a yellow experience just by learning about the relevant neuroscience. But this is absurd. No matter how much neuroscience Mary learns, she will never be able to know what it's like to have a color experience until she actually has one.* While she is confined to

* Haven't I contradicted myself? Earlier I said that the knowledge argument doesn't involve the claim that "If materialism is true, then anyone who knows enough neuroscience will be able to experience colors," whereas now I'm saying that if materialism is true, then "Mary . . . could come to know what it's like to have a yellow experience just by learning about the relevant neuroscience," which would be the case only if knowing the relevant neuroscience made you actually have color experiences. In fact, there is no contradiction here. The point is that *if* materialism were true, you'd be able to know what it's like to see yellow just from reading neuroscience *without* actually needing to have a yellow experience. But of course you cannot know what it's like to see yellow just from reading neuroscience without actually having a yellow experience; therefore, materialism is false.

her black-and-white room, Mary will have no grasp of (non-monochrome) color qualities and hence her knowledge of color experience will necessarily be incomplete.

This limitation on the expressive power of neuroscience entails a corresponding limit on its explanatory power. For if a neuroscientific theory were able to explain subjective qualities, it would have to be able to talk about them. Such a theory would start with a complete characterization of, say, the yellowness of yellow experiences, and then proceed to account for it in terms of more fundamental physical processes in the brain. But the language of neuroscience is simply not able to characterize subjective qualities; and if it can't even characterize them it certainly can't explain them.

Jackson's particular formulation focuses on the vivid case of color experience, but we could run similar arguments with respect to the qualities involved in other forms of subjective experience. We might imagine a deaf scientist learning about the neuroscience of musical experience, or someone who has never tasted Marmite learning what goes on in the brain when someone does. In each case, the purely quantitative vocabulary of the physical theory will be unable to convey the character of the subjective qualities involved in the relevant experiences. Galileo did not believe that the purely quantitative language of physical science could capture the qualities we find in our experience. The knowledge argument demonstrates that he was right after all.

BLUE BANANAS AND YELLOW TOMATOES

Daniel Dennett is one of the most radical and uncompromising of materialist philosophers. One of the "four horsemen" (together with Richard Dawkins, Sam Harris, and the late Christopher Hitchens), Dennett is a fearsome defender of atheism and materialism.* With his sturdy frame and lengthy beard, he strikes fear into the hearts of dualists everywhere.

Many scientists and philosophers conceive of the problem of consciousness as the challenge of working out how the brain produces the inner subjective world of consciousness. Dennett mockingly rejects this project as akin to wondering how the metaphysical reality of Hogwarts is produced by the letters on the pages of J. K. Rowling's novels. Consciousness, for Dennett, is a fiction conjured up by the information processing of the brain. The brain tricks us into thinking there is a magical inner world, just as the magician tricks us into thinking he has sawn a lady in half. It is an old joke by now that his best-selling book *Consciousness Explained* should really have been called *Consciousness Explained Away*.

I spent a week with Dennett on a tall-masted yacht in the Arctic. This was a conference-at-sea funded by the Russian investor and cofounder of the Center for Consciousness Studies Dmitry Volkov. Twelve philosophers and half a dozen graduate students from Moscow University spent seven days

* Despite their cultural association, atheism and materialism need not go hand in hand. Another of the "four horseman," Sam Harris, takes the problem of consciousness very seriously and is open to panpsychism as a solution.

and nights amidst the ice of Greenland wrestling with the topics of consciousness and free will, spotting whales and hiking on uninhabited islands along the way. Most of the philosophers were, like Dennett, committed to the view that consciousness is in some sense an illusion. But, to provide argumentative ballast, an official opposition party was invited onboard, composed of myself and two dualists: David Chalmers and Martine Nida-Rümelin. I can tell you now that we were given a hard time in discussion, although I like to think we gave as good as we got.

It was on this boat that I secured one of my proudest philosophical achievements: I managed to persuade Dennett he was wrong. Not about everything of course—philosophers are rarely persuaded to give up their comprehensive worldview—but about a reasonably significant aspect of the debate. In the last chapter we discussed the challenge to dualism of accounting for the causal interaction between the nonphysical mind and the physical brain. Both Dennett and the Churchlands (discussed in chapter 1, who were also on board) have argued that such interaction is ruled out by the principle of conservation of energy: the physical law that energy in an isolated physical system can be neither created nor destroyed.[9] Their reasoning is as follows: If the immaterial mind acted on the brain, then this would add energy to the brain that had not previously been there, thus violating the principle that energy in the physical universe is always conserved at the same level.

In the Q&A following Paul Churchland's talk, I suggested that this was not a good argument against dualism, as dualists could consistently hold that their psycho-physical laws (discussed in the last chapter) also respected the principle of conservation of energy. After all, according to our current best science, there are multiple laws of nature, which all work together to respect energy conservation. There seems no reason to think that we could not add more laws which also respect energy conservation, and no obvious reason they could not be psycho-physical laws.

I didn't expect the reaction that followed my comment. Russian graduate student Anton Kuznetsov later remarked to me that the other philosophers turned on me "Like a . . . [he consulted his friend for the correct English phrase] . . . a pack of wolves!" Patricia Churchland exclaimed, "So basically, you believe in magic!" I pointed out that I was in fact not myself sympathetic to dualism, precisely because of the difficulties making sense of mind-body interaction, and that I was merely making a technical point that this significant challenge for dualism has nothing to do with energy conservation.

I didn't get very far during the Q&A, but that evening Dennett and I stayed on board while the others took smaller boats to an island to explore. As we sat on deck, with Dennett carving a walking stick, I pushed the matter further, continually trying to hone down on the very specific point that conservation of energy is consistent with dualism. Eventually Dennett conceded, "Maybe that's right."

Despite this small concession with respect to the case *against* dualism, Dennett has zero time for the case *for* dualism. Like his fellow materialist Patricia Churchland, Dennett has little time for thought experiments, unless they are thought experiments intended to debunk the confused and oversimplistic intuitions that he believes drive most philosophical theories. As you might imagine, he is not prepared to learn lessons about consciousness from the far-fetched tale of Black and White Mary. While most materialists concede that the knowledge argument has some force, Dennett flat out denies its central claim: that Mary in her black-and-white room could not work out what it's like to see yellow. He puts his case as follows:

> It is of course true that in any realistic readily imaginable version of the story, Mary would come to learn something [when she leaves her black-and-white room], but in any realistic, readily imaginable version she might know a lot, but she would not know everything physical. Simply imagining that Mary knows a lot, and leaving it at that, is not a good way to figure out the implications of her having "all the physical information" any more than imagining she is filthy rich would be a good way to figure out the implications of the hypothesis that she owned everything.[10]

Dennett's response is initially quite persuasive. He plays on the incomprehensible enormity of knowing "all the physical facts," down to the tiniest details of each individual field or particle, encouraging the reader to think herself foolish for speculating about what would follow from such an unimaginable state of knowledge. Although this is how Jackson sets things up, in fact, one need not suppose that Mary knows *everything* physical to make the argument work.

In the materialist picture of the world, the physical sciences form a hierarchy, with higher-level sciences, such as neuroscience and cellular biology, explained in terms of more basic sciences, such as chemistry and physics. And if materialism is true, it is neuroscience, not fundamental physics, that will explain facts about the human mind. Basic physics will be the ultimate foundation of consciousness, in the sense that the electrochemical processes that make up experience will be ultimately constituted of the entities described by physics. But the facts of basic physics will be irrelevant to the immediate explanation of human consciousness.

Thus, we do not need to imagine that Mary knows all the physical facts, down to the exact number of up quarks and down quarks—which is fortunate, as no human being could process all that information!—we merely need to imagine that she has full knowledge of the account of color experience provided by complete and final neuroscience. If we appreciate this, the Mary thought experiment becomes a lot less far-fetched, and it becomes much more reasonable to think we can form a judgment about what she would or wouldn't know. All we are imagining is that she has more of the same kind of information available to contemporary neuroscientists.

Indeed, we can put the essential point of the knowledge argument without indulging in any imaginary scenarios whatsoever. All that is really needed to run the knowledge argument is the following premise:

> *The Key Premise*—Someone who has been blind from birth could not, through reading neuroscience (in braille), discover what it's like to have a yellow experience.

This is an extremely plausible premise. No matter how much a congenitally blind neuroscientist learns about the physical working of the brain, she will never be able to fully grasp the yellowness of a yellow experience.

Indeed, we can gain support for this premise by reflecting on real-life cases. Perhaps most interesting in this context is a paper on the knowledge argument written by Knut Nordby, who is about as close as you can get to a real-life Mary. Nordby is an expert in color vision who has *achromatopsia:* a rare condition in which, due to the absence of retinal cones, one is unable to perceive any colors, apart from black and white and shades of gray. In a sense, Nordby has spent his entire life in a black-and-white room. His reflections on the topic of Mary are fascinating:

> Can an achromatopic person ever have any idea what a colour experience is? Most achromatopic people think of colour as some curious property of surfaces that for them is somehow related to their apparent brightness. . . . To be able to cope in the world of colour-sighted people and avoid embarrassment, most achromatic people teach themselves the colours of common objects and the cultural "meaning" of some colours (e.g., red for danger, green for clear, blue for sadness).
>
> . . . One way for me to attempt to visualise the special quality of experiencing colour is to liken colour to the musical quality of tones, or *chroma*. Whereas colours have brightness and hue . . . tones have loudness, pitch, timbre and chroma. . . . Although the "chroma metaphor" may convey the idea of a special sensory property solely as an abstract thought exercise, it can never depict the actual experience of colours. Colours, like tones and tastes, are firsthand sensory experiences, and no amount of acquired theoretical knowledge can create this experience.[11]

It is clear that Nordby takes there to be some essential features of color experiences that he's missing out on. While he can, by comparison with sounds, form a kind of abstract template for the kind of qualities that characterize color experience, Nordby has no way of filling in that template. He lacks a grasp of the subjective quality itself. While he stops short of fully endorsing the knowledge argument's antimaterialist conclusion, he does concede the following:

> When, on her twenty-first birthday, she is let out into the world of full colors, will she be able to experience the color hues and identify them on the basis of her acquired knowledge?
>
> I believe that Mary will be able to sense and discriminate color hues but will not be able to name them on the basis of her knowledge.[12]

Dennett sees no reason even to give this much ground. With tongue firmly planted in cheek, he imagines his own ending to the Mary story:

> And so, one day, Mary's captors decided it was time for her to see colors. As a trick, they prepared a bright blue banana to present as her first color experience ever. Mary took one look at it and said, "Hey! You tried to trick me! Bananas are yellow, but this one is blue!" Her captors were dumbfounded. How did she do it? "Simple," she replied. "You have to remember that I know *everything*—absolutely everything—that can ever be known about the physical causes and effects of color vision. So of course before you brought the banana in, I had written down, in exquisite detail, exactly what physical impression a yellow object or a blue object (or a green object, etc.) would make on my nervous system. So I already know exactly what *thoughts* I would have. . . . I was not in the slightest surprised by my experience of blue (what surprised me was that you would try such a second-rate trick on

> me). I realize it is *hard to imagine* that I could know so much about my reactive dispositions that the way blue affected me came as no surprise. Of course it's hard for you to imagine. It's hard for anyone to imagine the consequences of someone knowing absolutely everything physical about anything!"[13]

Again, Dennett is playing on the immensity of being able to know absolutely everything about the physical processes going on in the brain. But as we have seen, the knowledge argument need not be set up in this way. All we need to imagine is that Mary has a working knowledge of the physical mechanisms underlying color experience; we don't yet have all the details, but we have made considerable progress and a complete account of these mechanisms will presumably be some nonrevolutionary extension of what we have already. The neuroscience we have already gives color-blind scientists like Nordby absolutely no grasp of the qualities of color experience, and it seems extremely unlikely that the completion of this ongoing project would make a substantial difference in this regard. Nordby already knows a great deal about the mechanics underlying color experience and is still utterly clueless as regards the yellowness of a yellow experience. Why would we suppose that adding some more details to the neuroscientific picture would suddenly make Nordby click his fingers and say, "Aha! So that's what it's like to have a yellow experience!"

(A more subtle point: Note that in Dennett's story, Mary learns what colors the objects are by working out what *thoughts* she would have when she has different color experiences. But if materialism is true, Mary ought to be able to work out what it's like to have a yellow experience simply by reflecting on the underlying neurological processes that constitute yellow experiences. Even Dennett doesn't explicitly endorse this absurd implication of materialism.)

In fact, Dennett is not really trying to positively defend the view that Mary *would* be able to work out what it's like to see yellow, he simply wants to undermine his opponent's claim that she *wouldn't:*

> Can you prove it? My point is not that my way of telling the rest of the story proves that Mary *doesn't* learn anything, but that the usual way of imagining the story doesn't *prove* that she *does.*

Science and philosophy rarely prove anything with 100 percent certainty. We can't prove with 100 percent certainty that we're not in the Matrix being fed a virtual reality by evil computers. We can't prove, as Bertrand Russell famously pointed out, that the world didn't spontaneously come into existence five minutes ago, including from its start all of our memories suggesting a history that never took place. What these examples show is that we shouldn't ask for certainty, but rather for beliefs which are reasonably supported by the evidence.

The great progress of neuroscience in understanding the mechanisms underlying color experience has been unable to provide the color-blind with *any* insight—zero, nothing, nada—into what it's like to have color experience, and there is no reason at all to think this will change when a few more details are added.* On this basis, the proponent of the knowledge

* Of course, third-person science can record the *structural* features of experiences, such as the commonalities between color experiences in terms

argument can believe with confidence the key premise of the argument: that neuroscience cannot teach the blind/color-blind what it's like to have color experience. We never know for sure what the future will bring, but as the physicist Wolfgang Pauli used to put it, science is not in the business of claiming "credits for the future." In other words, we can only go off the evidence we currently have, and the evidence we currently have strongly suggests that the key premise of the knowledge argument is true. Dennett is applying double standards: artificially ramping up the onus of proof the antimaterialist is expected to meet to a ridiculously high standard that not even our best science is expected to meet.

Many philosophers and scientists start from the position that the truth of materialism is overwhelmingly likely: better to embrace some of its more improbable consequences given that the alternatives are even more improbable. But as we discovered in the first chapter, the main motivation for embracing materialism arises from an incorrect view of the history of science. It is thought that the great and rightly celebrated success of physical science gives us overwhelming reason to embrace materialism as the true theory of consciousness (as well as everything else). In fact, physical sciences have been so successful precisely because from Galileo onward they set aside the *qualitative* in order to focus on the *quantitative*. The fact that physical science has done well when it ignores consciousness gives us no reason to think it will do well when it tries to apply its quantitative methods to consciousness itself.

---

of their brightness, hue, and saturation. It is this kind of information that allows Nordby to get a grip on the structure of color experience. What third-person science cannot convey are the subjective qualities that realize these structural features.

THE ZOMBIE THREAT TO A SCIENCE OF MIND

We started this chapter and the last staring at the brain of your best friend Susan, wondering if there's an invisible immaterial mind lurking within. In the absence of positive evidence, the materialist is inclined to reject what cannot be seen or otherwise perceived with the senses. But this quite sensible demand for observational evidence can lead us into trouble when it comes to consciousness, as consciousness is itself unobservable. I know I am conscious due to my immediate awareness of my own feelings and experiences. But as I stare into the complex workings of Susan's brain, how do I know whether *she* is conscious?

Even accepting that consciousness cannot be directly perceived in Susan's brain, you might think that it's manifest in her behavior. Imagine your last conversation with Susan. Perhaps she told you how much fun she had had on her recent holiday and smiled as she flicked through the photos on her phone. Surely her smile is evidence that she's enjoying herself? Or maybe she told you that her pet lizard Nigel passed away and welled up a little as she recounted the sad news. Surely her tears are good evidence that Susan is feeling sad?

Such inferences from behavior to consciousness are standard in daily life. But how well grounded are they? If materialism is true, then Susan's macroscopic behavior is ultimately determined by the quarks and electrons that make her up, acting in accordance with the basic laws of physics. And hence she would behave the same—smile real smiles, cry real tears—whether or not she is conscious. How then can you rule out the possibility that Susan is just a complicated mechanism disposed to behave *as though* she had feelings and experiences?

To introduce a bit of philosophical jargon, we are currently discussing the issue of how we know whether or not Susan is a *zombie.* The word "zombie" is a technical term in philosophy for a certain kind of imaginary creature. Having introduced that word, it is crucial to distinguish sharply between *philosophical* zombies and *Hollywood* zombies. It would be quite easy to tell if Susan were a Hollywood zombie, as she'd look something like this:

Whereas if Susan were a philosophical zombie, she'd look like this:

In other words, philosophical zombies look just like ordinary women and men. And they don't wander around, arms outstretched seeking the flesh of the living. Rather, philosophical zombies walk, and talk, and in all ways behave just like ordinary human beings. And the reason a philosophical zombie behaves just like an ordinary human being is that the physical workings of its body and brain are indiscernible from those of an ordinary person.

But there's a crucial difference: a philosophical zombie, by definition, is not conscious. If you stick a knife in a philosophical zombie, it'll scream and try to get away, but it doesn't actually *feel* pain. When it crosses the road, a philosophical zombie looks both ways, waits for the traffic to die down, and then carefully crosses the road, but it doesn't actually have any visual or auditory *experience* of the road around it. A philosophical zombie is just a complicated mechanism set up to behave like an ordinary human being.[14]

How can you possibly know whether or not Susan is a zombie? By the very definition of a zombie, Susan would behave exactly the same if she was one. This is the philosophical challenge known as *the problem of other minds.* One purpose of thinking about zombies is that it helps make the problem of other minds vivid.

Historically, one popular solution to the problem of other minds is to offer an argument from analogy. I know in my own case that certain physical states are correlated with certain conscious experiences (or at least I could in principle find out by fiddling with my own brain and noting what I experience). I can then reason as follows: Other people have similar things going on in their brains, therefore, most likely they have similar experiences to mine. This strategy relies on a generalization from my own case to the case of everybody else. But once explicitly pointed out, this looks like an illegitimate inference: knowing that something holds in a single case does not usually allow you to infer that it holds in all cases. Suppose I'd only ever seen one white swan. I could not infer on that basis that all swans are white. What we need to know is whether my case is representative of human beings in general. But if we knew this, we wouldn't have a problem of other minds in the first place.

Philosophy is filled with these kinds of skeptical arguments:

How do we know there's an external world? How do we know the future will resemble the past? How do I know my memories can be trusted? Despite the best efforts of philosophers, none of them seems to have a solution. Kant thought it was a great scandal that philosophers had still not even managed to prove the existence of the external world. Personally, I'm not so bothered by the skeptical doubts of philosophers. I'm inclined to think that there are some things we just have to take as basic, if only because we need to get on with life.* The fact that other people have minds is probably one example of this.

However, that's not the end of the philosophical interest in zombies. The fact that you cannot rule out for certain the possibility that Susan is a zombie shows something interesting. (Warning: Things are about to get a little complicated. The reader can skip to the next section if you are already persuaded that physical science cannot account for consciousness and want to be spared the details of a complex logical proof. . . .) What this shows is *that there's no contradiction in the idea that something with the same physical nature as Susan could lack an inner subjective life.* For if certain facts about Susan's body and brain were inconsistent with her being a zombie, then you could rely on this fact to prove that Susan is not a zombie. The very fact that there is a problem of other minds entails that zombies are logically possible, in the sense that there is no inconsistency or incoherence in the idea of a zombie.

We could put the argument as follows:

* John Locke perhaps put it best when he said, "He that in the ordinary affairs of life, would admit of nothing but direct plain demonstration, would be sure of nothing in this world but of perishing quickly." (Taken from book IV, chapter XI, section 10 of Locke's *An Essay Concerning Human Understanding*.) In other words, it's impossible to live out the skeptical doubts of philosophers.

The Argument for the Logical
Possibility of Zombies

1. If zombies were logically impossible, I'd be able to prove that Susan is not a zombie.
2. I cannot prove that Susan is not a zombie.
3. Therefore, zombies are logically possible.

To preempt potential misunderstanding, I am certainly not saying that it's *reasonable* for you to think that Susan is a zombie. We of course assume that creatures that behave in ways similar to ourselves have consciousness similar to our own, and it's completely reasonable to do so, if only because life would be intolerable if we succumbed to the skeptical doubts of philosophers. All that the above argument shows is that zombies are logically possible.

In order to make the point clear, contrast flying pigs and square circles. Neither of these things exist. But flying pigs have a slight advantage over square circles in that they are at least logically possible: there is no contradiction in the idea of a flying pig. We all know of course that there are no flying pigs, and perhaps in our universe such things are impossible. But if things had been very different, perhaps if gravity had been slightly weaker and pigs had evolved with wings, there could have been pigs hurtling through our skies. In contrast, no matter how weird or wacky our universe and its laws of nature had turned out, there couldn't have been square circles. And that's because the very idea of a square circle is contradictory.

What the insolubility of the problem of other minds reveals is that philosophical zombies are more like flying pigs than square circles. Nobody thinks that philosophical zombies exist, any more than they think flying pigs exist. But there is no contradiction in the idea of a zombie, and hence if our universe had been very different, perhaps if the laws of nature had been different, there could have been zombies roaming our planet.

You might still be unimpressed. Who cares what's *possible*? All manner of things are possible—there's no contradiction, after all, in the idea of a trillion angels dancing on the head of a pin—but what most of us care about is what's *real*. However, there is something very important here. Just as Galileo's thought experiment demonstrated a contradiction in a central plank of Aristotelian physics, so philosophical zombies demonstrate a contradiction in materialism. It can be logically demonstrated that if zombies are even *possible*—not actual, merely logically possible—then materialism cannot possibly be true.

How could this be? How can a claim about *mere possibility* entail something about the real world? The argument hangs on a logical principle that is almost universally accepted by philosophers and logicians:

> *The Identity Principle*—If X and Y are identical, then it is logically impossible for X to exist without Y (or vice versa).*

* Academic philosophers may be concerned that I am failing to distinguish *logical* possibility from the supposedly broader notion of *metaphysical* possibility. I argue at length in my academic book *Consciousness and Fundamental Reality* (and in my short article "Essentialist Modal Rationalism," which builds on what I wrote in the book) that metaphysical possibility is just *logical possibility when you have a full understanding of what you're conceiving of.* I discuss a little in Technical Appendix A how the discussion here connects with the broader discussions in the academic literature (in particular, see footnote on p. 108). I hope the reader will forgive me if I don't get into the full technical details here so as not to compromise the accessibility of the text. But to preempt another concern: I am, of course, assuming that the terms in the identity statement are rigid designators.

To see the identity principle in action, let us consider the case of Charles Dodgson.

Charles Dodgson was a nineteenth-century logician at Oxford University. But you might know him by another name. Charles Dodgson was also Lewis Carroll, the author of *Alice's Adventures in Wonderland* and *Alice Through the Looking-Glass.* Now the identity principle tells us that because Charles Dodgson and Lewis Carroll are identical, they couldn't possibly exist apart. You couldn't have Dodgson in one room and Carroll in the other. Why not? Because there aren't two people that could be separated: Charles Dodgson is one and the same person as Lewis Carroll. Not even an all-powerful being could tear apart Charles Dodgson from Lewis Carroll.*

What has this got to do with materialism? I get bored very easily at social events, and so I often try to engage the nearest person in a philosophical discussion. Sometimes this is welcome, although in an equal number of cases I sense my collocutor waiting for the most convenient opportunity to politely excuse him- or herself. In any case, I have discovered via such discussions that many people are under a profound misapprehension as to what materialism is. Many people take materialism to be the view that the brain *produces* consciousness, as though consciousness were some peculiar kind of gas that the physical workings of the brain bring into being. However, such a view would not be materialism, as it implies that consciousness is

* It's of course possible that someone else might have written *Alice's Adventures in Wonderland* and *Alice Through the Looking-Glass.* But by "Lewis Carroll" we don't mean "whoever wrote the Alice books." If that were the case, then the statement "It turns out that it was Lewis Carroll's sister who really wrote the Alice books" would be logically contradictory, which is clearly not the case (that sentence may be false, but it's not contradictory). This is a point made (with a different example) by Saul Kripke in his book *Naming and Necessity.*

something over and above the physical workings of the brain. Compare: my parents *produced* me, and as such I am a separate entity from my parents. Similarly, if consciousness were *produced* by the brain, then consciousness would be something separate and distinct from the physical workings of the brain, just as a child is separate and distinct from its parents.

In fact, materialism is the view that feelings and experiences are *identical* with states of the brain. Materialists do not believe that experiences are *caused* by brain states—for in that case experiences would be separate and distinct from brain states. Rather, materialists believe that experiences *just are* brain states: experiences and brain states are *one and the same thing*. For the materialist, physical science tells us what experiences really are—electrochemical processes in the brain—just as chemistry tells us what water/lightning really are—H_2O/electrostatic discharge.*

It is important to absorb this in order to really grasp what materialism amounts to. Imagine that it is you, rather than your friend Susan, who has had the top of your head removed, and suppose that a neuroscientist is peering into your head and observing your brain states. If materialism is true, what the scientist is looking at are not the states that *produce* your experiences, but rather the experiences themselves. Those very feelings and experiences you know "from the inside," and the brain states the scientist sees "from the outside," are one and the same thing seen from two different perspectives.

We are now in a position to understand why the mere possibility of zombies is inconsistent with materialism. For mate-

* Some materialists insist that the relationship between conscious states and physical states is one of *constitution* rather than *identity*. However, there is almost universal agreement among academic philosophers (a rare thing!) that the possibility of zombies is equally inconsistent with this view, and so, for the sake of simplicity, I will set it aside.

rialism tells us that feelings and brain states are identical. But according to the identity principle, if feelings and brain states are identical, then they couldn't possibly exist apart. If a certain kind of activity in the hypothalamus (a part of the brain involved in regulating appetite) is identical with the feeling of hunger, then that kind of hypothalamus activity could not possibly exist in the absence of the feeling of hunger, just as Charles Dodgson couldn't possibly exist without Lewis Carroll, or water without H_2O. And yet that is precisely what happens in a philosophical zombie. Your zombie twin has all of your physical brain states but lacks your feelings and experiences; when denied food there is the corresponding kind of activity in its hypothalamus but it doesn't actually feel hunger. It follows that the mere possibility of zombies is inconsistent with materialism.

We can put the argument as follows:

The Zombie Argument

1. If materialism is true, then feelings are identical with brain states.
2. If feelings are identical with brain states, then it is not logically possible for feelings to exist without brain states, or vice versa. (This follows from the identity principle.)
3. If zombies are logically possible, then it is logically possible for brain states to exist without feelings.
4. Therefore, if zombies are logically possible, materialism is false.
5. Zombies are logically possible (as demonstrated by the argument on p. 90).
6. Therefore, materialism is false.

The reader may feel like a trick has been pulled on her. But all we are doing is carefully drawing out the logical implications of materialism, just as Galileo drew out the logical implications of Aristotelian physics.

The zombie argument is generally known in the academic philosophical literature as the "conceivability argument." I think this is something of a misnomer, as it suggests that the argument has something to do with what can be imagined. This has led to some misunderstanding, as can be seen, for example, in Anil Seth's response to the argument:

> Conceivability arguments are generally weak since they often rest on failures of imagination or knowledge, rather than on insights into necessity. For example: the more I know about aerodynamics, the less I can imagine a 787 Dreamliner flying backwards. It cannot be done and such a thing is only "conceivable" through ignorance about how wings work.[15]

There are two problems with what Seth says here. First of all, the zombie argument is not concerned with what can be imagined—as though the limits of human imagination were a guide to the limits of reality—but rather about what is logically possible, in the sense of being free from contradiction. Secondly, once we appreciate that it is logical possibility that is the focus, we can see that Seth's analogy is not to the point, as it involves a different sense of "possibility."

The word "possible" is used in (at least) two senses:

Logical possibility—Something is logically possible if it is free from contradiction.

Natural possibility—Something is naturally possible if it is consistent with the laws of nature.

The zombie argument is concerned with logical possibility, whereas Seth's example deals with natural possibility. It is inconsistent with the laws of nature for a 787 Dreamliner to fly backward, and one appreciates this as one learns about the relevant laws of nature. But it is certainly not *contradictory* for a 787 Dreamliner to fly backward; if the laws of nature had been very different, such a thing might have been possible. In other words, a 787 Dreamliner flying backward is not *naturally* possible but it is *logically* possible.

In a way, the zombie argument is a long-winded way of demonstrating something that's pretty obvious. To describe a person in terms of his or her physical nature is to ascribe to that person certain *objective, quantitative* properties. To describe someone in terms of his or her conscious experiences is to ascribe to that person certain *subjective, qualitative* properties. And the fact that a physical system has the former kind of properties does not entail that it has the latter kind of properties. This is the core of the problem of consciousness. Contemporary materialism is not a solution but a stubborn refusal to face up to the problem.

IS CONSCIOUSNESS AN ILLUSION?

Keith Frankish is a brilliant philosopher. He's also one of the warmest people I know. And I know from a friendship of a few years now that Keith has a real empathy for human beings as well as a deep concern for the state of the world.

In spite of this, Keith doesn't believe in consciousness. He doesn't believe that anyone has ever felt pain, or seen red, or tasted chocolate. At least not if we are understanding these experiences, as we have throughout this book, as involving subjective qualities. And it's not that Keith thinks that subjective qualities are an artificial creation of philosophers; he agrees that it's very natural for us to think of ourselves as having experiences involving subjective qualities. It's just that he thinks it's an illusion, a trick played on us by our brains. Consciousness is no more real than fairy dust.

Why on earth does Keith think this? Basically, because he agrees with the kind of arguments presented in this chapter. It is as clear to Keith as it is to me that the subjective/qualitative properties of consciousness cannot be accounted for in the objective/quantitative language of physical science. However, from this shared starting point, Keith and I move in polar opposite directions. In my view, the fact that physical science can't explain consciousness shows that we need a new scientific paradigm, one that is able to accommodate the reality of consciousness (see next chapter). Keith agrees that physical science cannot explain consciousness but infers from this that consciousness must be an illusion.

Here's how he puts it:

> Suppose we encounter something that seems anomalous, in the sense of being radically inexplicable within our established scientific worldview. Psychokinesis is an example. . . . we could accept that the phenomenon is real and explore the implications of its existence, proposing major revisions or extensions to our science, perhaps amounting to a paradigm shift. In the case of psychokinesis, we might posit previously unknown psychic forces and embark on a major revision of physics to accommodate them . . . or we could argue that the phenomenon is illusory and set

> about investigating how the illusion is produced. Thus, we might argue that people who seem to have psychokinetic powers are employing some trick to make it seem as if they are mentally influencing objects.[16]

Frankish hopes the scientifically inclined reader will favor the second option with respect to psychokinesis. One should explore every option for explaining away psychokinesis as an illusion before we start reshaping science to accommodate it. But if this is the rational approach to take with psychokinesis, why not also with consciousness? If the arguments of this chapter are sound, then consciousness is at least as ill-fitting with our current scientific paradigm as psychokinesis. And after all, the only evidence we have for the existence of consciousness is that it *seems to people* that they are conscious. If we can explain away that seeming as an illusion, and thus avoid a radical expansion of our current scientific paradigm, then surely that should be the preferred option.

This is a beautiful, elegant solution to the problem of consciousness that easily avoids all of the difficulties raised thus far in this chapter. If consciousness does not exist, then we don't need a scientific theory of it, any more than we need a scientific theory of astrology or alchemy. This is not to say the view is problem-free. The central challenge for *illusionism*—as Frankish has dubbed his position—is to explain how the brain manages to pull off this remarkable magic trick, how it manages to convince us to believe with such certainty that subjective qualities really exist. But solving this form of the "problem of consciousness" does not require a radical revision to our current scientific paradigm.

You may remember from a few pages ago that I like to bombard unwilling members of the public with philosophical discussion, in order to avoid the monotony of, say, a family wedding reception. So long as it's before dessert, I like to see how people react to the illusionist position. Probably the most common reaction (apart from a stifled yawn and a glazed expression) is, "That's impossible! If there is no consciousness, what is the 'I' that thinks it's conscious??? *Who* is it that is subject to this illusion?"

Philosopher Galen Strawson nicely articulates what I think lies behind this objection to illusionism—a view which Strawson mockingly calls "the silliest claim ever made":

> One of the strangest things the Deniers [Strawson's provocative term for illusionists] say is that although it seems that there is conscious experience, there isn't really any conscious experience: the seeming is, in fact, an illusion. The trouble with this is that any such illusion is already and necessarily an actual instance of the thing said to be an illusion. Suppose you're hypnotized to feel intense pain. Someone may say that you're not really in pain, that the pain is illusory, because you haven't really suffered any bodily damage. But to seem to feel pain is to be in pain. It's not possible here to open up a gap between appearance and reality, between what is and what seems.[17]

You might think that as an opponent of materialism I'd be keen on embracing this quick and dirty rejection of illusionism. Unfortunately, things are not so simple (they rarely are . . .). Whether or not illusionism is coherent depends on the thorny question of whether computers can think. It's time for another extended digression.

CAN COMPUTERS THINK?

We have met a lot of "fathers" in this book. In chapter 1 we met both Galileo—the father of modern science—and Descartes—the father of modern philosophy. It's now time to meet the father of modern computing: Alan Turing (1912–1954). Turing was an extraordinary individual. During the Second World War he worked at Britain's code-breaking center, Bletchley Park, where he played a pivotal role in cracking the Germans' Enigma code. According to some estimates, this work shortened the war by more than two years, saving over fourteen million lives. Britain repaid Turing by prosecuting him for homosexual acts in 1952, for which he accepted chemical castration rather than prison as punishment. He died two years later from cyanide poising, which an inquest determined to be suicide.

Turing laid the ground for modern computing by rigorously formulating the notion of "computation," and testing through logical arguments its potential as well as its limits. Very roughly, a task is "computable" if it is possible to specify a sequence of instructions that will result in the completion of the task when carried out by some machine. We call the set of instructions an "algorithm." Turing was able to demonstrate that there are mathematical functions on the natural numbers that cannot be computed. Nonetheless, he remained hopeful that all of the functions of the human mind might be computable, which for many opens up the possibility that the mind itself might be a kind of computer.

One of most famous concepts in artificial intelligence is what has become known as "the Turing Test." Originally known as "the Imitation Game," this was Turing's test for the capacity of a machine to demonstrate intelligent behavior. Turing imagined a person—"the interrogator"—addressing a series of questions to two collocutors hidden from sight in an adjacent room, one of whom is another person and one of whom is a machine. The aim of the game is to work out which is which. To pass the test, the machine would have to fool 70 percent of judges during a five-minute conversation. Turing predicted that by the end of the twentieth century there would be machines able to pass this test with ease.

Turing's prediction was not borne out. Despite occasional media hype to the contrary, no computer has ever passed the Turing Test. This doesn't in itself show that there's anything wrong with the test, only that Turing was overly optimistic about the speed of technological advancement. But what exactly is it a test of? Suppose there is one day a computer that is able to pass the Turing Test, and can talk fluently about, say, the ethical and political issues raised by a globalized economy. Would we then want to say that the computer genuinely *understands* these social issues? Or is it merely parroting words *as though* it understands?

Philosophers are torn on this issue. The American philosopher John Searle designed a thought experiment intended to show that mere computation, even computation powerful enough to pass the Turing Test, is not sufficient for genuine understanding. This is the famous "Chinese Room" thought experiment.[18] Searle imagines a room containing a human being who doesn't speak Chinese, but who has a huge number of pieces of paper containing Chinese symbols and a book containing instructions telling her which Chinese symbol is correct to "output" given a certain Chinese symbol as "input." Outside the room there are native Chinese speakers, who pass questions in

Chinese under the door. The non–Chinese speaker inside the room receives these "inputs," looks them up in the book of instructions, and then delivers the correct "outputs."

If the book of instructions is detailed and wide-ranging enough, then the room will simulate the responses of a native Chinese speaker. And yet the person inside the room does not speak Chinese but is rather just blindly following instructions. And it is intuitively obvious, according to Searle, that the room itself does not understand Chinese. What we have is merely the appearance of understanding in the absence of genuine comprehension.

What Searle has done is provide a vivid way of reflecting on what a computer is. In line with how Turing defined "computation," a computer is something that follows instructions. The Chinese Room is effectively a kind of computer following a program, the program being the set of instructions contained in the book. If the program is good enough, the room's responses would be indistinguishable from those of a native Chinese speaker and hence will pass the Turing Test. And yet, once we see that all that is really going on is blind following of instructions, it is just obvious—according to Searle—that this is not sufficient for genuine understanding.

Many dispute Searle's argument. Note that, in contrast to the other thought experiments we have considered in this chapter, Searle has not demonstrated a *contradiction* in his opponent's position. The aim of the thought experiment is simply to make vivid what is involved in computation, with the hope that Searle's audience feels disinclined to call this "understanding."

Whether or not one reacts in the way Searle hopes may depend on how one defines the word "understanding." As Humpty Dumpty wisely observed, one is free to define words how one chooses. It is clearly an option to give the word "understanding" a computational meaning so that it's true to say that a machine that passes the Turing Test "understands." Indeed, this is exactly what Turing himself does. In his classic paper, "Computing Machinery and Intelligence," in which he formulates the Turing Test, Turing dismisses the question of whether or not a computer "understands" or "thinks" as hopelessly vague, before going on to *replace* our everyday definition of "thinking" with what he considers to be a more precise definition of "thinking." In other words, Turing simply *defines* the word "thinking" to mean passing his test (i.e., fooling 70 percent of judges in a five-minute conversation). If one accepts Turing's computational definition of thought, then it follows trivially that computers that can pass the Turing Test can think. If one follows Searle in rejecting Turing's definition, then the opposite conclusion may follow.

BACK TO ILLUSIONISM . . .

What has all this got to do with illusionism? It is crucial to note that computation has nothing to do with consciousness. To compute is simply to follow instructions. In Searle's thought experiment, a human being is manipulating symbols in the Chinese Room, but we could just as easily imagine a nonconscious mechanism programmed to follow the instructions. This is not to say that machines will never be conscious. If human brains can be conscious, then there is no reason to think that a man-made object could not also be conscious (the theory I will defend in the next chapter is perfectly consistent with this possibility). Even so, my point here is simply that computation does not *require* consciousness. A computer does not need to be conscious in order to run its program.

Let us suppose, for the sake of discussion, that Searle is wrong, and that the computers of the future will be able to think and understand, provided they can pass the Turing Test. If such a computer can converse on the global economy and express the need for a Keynesian approach, then we can truly say that that computer *believes* that a Keynesian approach is called for. Let us further suppose that the computers of the future will not be conscious. (Indeed, as we shall see in the next chapter, according to the Integrated Information Theory of consciousness, the processing of computers is not integrated enough to give rise to consciousness.) In this case, the computers in question will be able think and understand despite having no conscious experience at all.

We are imagining computers that have no feelings or experiences. But might such computers not be programmed to *believe* that they have feelings and experiences? It is hard to see why not. If a computer can be programmed to discuss in great detail the fate of the global economy, then it can be programmed to expound on the subtle and nuanced subjective experiences which—according to the computer—are part of its immediate awareness. We can imagine the droll intonations of the computer HAL from *2001: A Space Odyssey* calmly reporting, "I've been feeling slightly anxious this morning, Dave . . . but since you've been ignoring me, this slight anxiety has grown into a fierce anger. . . ." Of course, as HAL is a computer, his speech outputs are merely the result of mechanistically following instructions. But if we accept Turing's definition of "thinking," then, so long as HAL could pass the Turing Test, he/she/it will nonetheless count as a genuine thinker.

We saw above Galen Strawson arguing that illusionism is self-refuting on the grounds that "seeming" implies consciousness: If it *seems* to me that I'm in pain, then I'm pain. Whether or not this is true depends on what we mean by "seeming." The computer described above *thinks* (at least in the computational sense of "thinking" defined above) that it has feelings and experiences. Is this not a reasonable sense in which it "seems" to the computer that it is conscious? Indeed, this opens up the possibility that *we* might think (in the computational sense) that we're conscious even though we're not. Perhaps, indeed, the human mind is nothing more than a complicated computer programmed by evolution to think that it has subjective inner states.

Why would evolution want us to believe in things that don't exist? The psychologist Nicholas Humphrey, also a proponent of illusionism, has argued that the illusion of consciousness comes with a significant survival advantage.[19] Creatures who believe they have a subjective inner life develop a great interest in preserving and enriching that inner life through complex engagement with their environment. Ultimately, creatures who believe they are conscious come to believe in another illusion—the Self—the existence of which they are driven to preserve at all costs. I once gave a passionate talk about consciousness with Nicholas Humphrey in the audience. He was delighted by my performance, and in the Q&A remarked that I had in fact confirmed his own view, by demonstrating what passions the belief in consciousness is apt to bring about!

In fact, I have much more time for illusionism than I do for more standard forms of materialism that try to have their cake and eat it too. Most materialists want to assert the full-blown reality of subjective, quality-rich consciousness, while at the same time holding that reality is purely physical. But as we have seen in this chapter, it is demonstrably contradictory to hold both that the world can be exhaustively characterized in objective/quantitative terms and that it contains the subjective qualities of experience. Illusionists don't try to square this circle: in order to preserve a pristinely objective world, they reject subjectivity as an illusion.

However, although the view is coherent, I would argue that there is no reason to accept it and plenty of reasons for rejecting it. The proponents of illusionism claim there is scientific evidence in support of the view. However, scientific evidence is nothing other than *facts about experience*. I know there's a table in front of me in virtue of having *an experience* of the table. We know about electrons in virtue of *experiencing* vapor trails in cloud chambers. We have no way of directly accessing the physical world; all knowledge of physical reality is mediated through experience. Thus, the claim that there is scientific evidence that consciousness doesn't exist is self-defeating: in accepting illusionism one thereby undermines the (supposed) evidence for being an illusionist. It's a bit like believing that somebody never tells the truth because they tell you so.

Moreover, to put my cards on the table, I am on the side of Searle in the above debate. I would say that I know I have thoughts in a more than computational sense. And I know this because I am immediately aware of my thoughts as they arise in consciousness. As I lie back and observe the flow of my consciousness, I am aware not only of feelings and sensations, but also of *conscious thoughts:* wonderings and musings and fleeting recollections. Turing demanded to know what "thinking" means if not computation. I reply that the answer is obvious: the phenomenon of conscious thought that each of us enjoys every second of waking life. It is precisely this that the Chinese Room and a nonconscious computer lack.

We are used to thinking of consciousness in sensory terms: pain, seeing red, tasting citrus. But the reality of what we might call *cognitive consciousness*—conscious episodes of thought—is no less apparent to introspection. And when conscious thought is taken into account, illusionism is simply incoherent. A creature cannot consciously think it's conscious without thereby being conscious. To modify Descartes: I consciously think, therefore illusionism is false.

The final problem with illusionism is that the view is utterly unmotivated. There is arguably scientific evidence against dualism. But dualism is not the only alternative to materialism. There is a view that can perfectly well accommodate both the qualitative facts of subjective experience and the quantitative facts of physical science. This is the view we will explore in the next chapter.

Enough of the problems. Let's get some solutions.

TECHNICAL APPENDIX A: DO WE NEED TO EXPLAIN CONSCIOUSNESS?

I have been assuming in the above that the aim of materialists is to *explain* consciousness. Against this assumption, the materialist would need to give an account of how the subjective qualities of consciousness arise from the chemistry of the brain, in something like the way the liquidity of water arises from its chemical structure. The arguments in the chapter aimed to show that the resources of physical science are not up to this task.

However, in academic philosophy, the dominant view amongst materialists is that consciousness does not need to be explained. According to what we might call the "brute identity theory," conscious states are identical to brain states and that's all that needs to be said. We don't need to give some "explanation" of how, say, the feeling of pain can be accounted for in terms of a certain kind of brain state. If we have sufficient empirical evidence that pain is identical with brain state X, which brute identity theorists claim we would have if pain were found to be systematically correlated with brain state X, then we can simply assert the identity and the case is closed. It is a philosophical confusion, according to the brute identity theory, to suppose that anything more is required.*

The brute identity theory is very unsatisfying. Science is supposed to offer explanations. I want to know *how* processes in the brain result in a subjective inner world of feeling and experience. To simply be told that it's a brute identity and there's nothing more to be said is a bit like the "Because I said so" answer a parent gives to the incessant "Why?" questions of a toddler. Moreover, in the case of other scientific identities we do get an explanation of the manifest features of the phenomenon in question. When lightning is identified as a certain kind of electrostatic discharge, we get an explanation of its appearance and of the thundering sound that follows it; when water is identified with H_2O, we get an explanation of its liquidity and its boiling point; when heat is identified with molecular motion, we get an explanation of its effect on our bodies. But when pain is identified with a certain kind of brain state, in contrast, we don't get a satisfying explanation of why pain *feels* the way it

* But doesn't the zombie argument demonstrate that brain states can't be identical with feelings? Brute identity theorists respond to the zombie argument by distinguishing *logical possibility* from the supposedly broader notion of *metaphysical possibility*. They then interpret the identity principle in terms of metaphysical, rather than logical, possibility. The result is that the identity between experiences and brain states is compatible with the logical possibility of zombies, and premise 2 of the zombie argument is false. I respond to this at length in *Consciousness and Fundamental Reality*, arguing that metaphysical possibility is just logical possibility when you know what you're conceiving of.

does. As the philosopher Joe Levine famously put it, there is an "explanatory gap" involved in mind-brain identities that we don't find in the case of other scientific identities.[20]

There is another crucial difference between standard scientific identities and putative identities between experiences and brain states. In any standard scientific identity, we begin by thinking about a given phenomenon *indirectly* in terms of its superficial characteristics. We think of water in terms of its being colorless, odorless, filling oceans and lakes, etc.; we think of gold in terms of its yellowish appearance. Then scientists come along and tell us the essential nature that lies behind those superficial characteristics: we discover that water is H_2O and that gold is the element with atomic number 79. Contrast this with the case of pain. In ordinary thought, we conceive of pain in terms of how it feels. But pain *just is* a feeling, and, by definition, there's nothing more to a feeling than how it feels. Insofar as we're thinking of *the feeling of pain considered in and of itself,* ordinary thought already tells us what it is. The job of science, then, is not to tell us what feelings are (we already know what a feeling is when we feel it) but rather to give an account of the place of feelings in a general theory of reality.*

This is a big debate, and these few paragraphs do justice neither to the brute identity theory (known in the academic literature as the "phenomenal concept strategy") nor to my response to it. You can find much more detail in my academic book *Consciousness and Fundamental Reality.*

* The philosopher Saul Kripke emphasized something like this distinction between standard scientific identities and mental/physical identities in his book *Naming and Necessity.* I have developed the idea in detail in *Consciousness and Fundamental Reality.*

4

How to Solve the Problem of Consciousness

When I was a philosophy undergraduate, in the dying embers of the twentieth century, we were taught that there were only two options for dealing with consciousness. Either you turned to physical science to explain consciousness, in which case you were a materialist, or you thought consciousness existed outside of the physical realm altogether, in which case you were a dualist. As far back as I can remember I have been obsessed with the problem of consciousness, and so from day one of student life I decided to read everything I possibly could on these two "live options" for explaining consciousness.

In my first year of university, I decided I was a committed materialist. The wealth of scientific support for placing consciousness in the brain seemed to rule out any other option. I passionately debated religious dualists in support of the identity of the mind and the brain, declaring that any other option lacked scientific credibility.

But the more I thought about the arguments, the more I came to doubt the coherence of conventional materialism. The picture of the universe painted by physical science seemed to have no place for the subjective qualities of experience. I decided an uncompromising rejection of the reality of consciousness was the only consistent materialist position.

Sadly, pretending you're not conscious is a tough act to pull off, and as my second year wore on, I increasingly felt a peculiar kind of cognitive dissonance. I began to feel dishonest when presenting my case in discussions. Even alone I felt a certain insincerity in myself, an uncomfortable case of what the existentialists called "bad faith."

It all came to a head one evening as I sat in a noisy crowded bar, feeling the banging beat of the music in my chest as I enjoyed the taste of lager and the rush of nicotine from my first cigarette of the evening (smoking in bars was legal in those days), and I was suddenly overcome with a vivid sense of how real these conscious experiences were, and of their clash with my official worldview. I pushed my way out of the bar and stood in the cold rain with my eyes closed. I couldn't deny it anymore. I'd already accepted that if materialism was true, then I was a zombie. But I knew I wasn't a zombie; I was a thinking, feeling human being. I could no longer live in denial of my consciousness.

After that transformative experience I became something of a closet dualist. The weight of the scientific case against dualism still bothered me deeply, and—a little unfairly in hindsight—I associated dualism with the religious upbringing I had rejected at the age of fourteen by refusing to get confirmed Catholic, much to the consternation of my matriarchal grandmother. In my final-year dissertation, I reluctantly argued that the problem of consciousness could not be resolved. Feeling disappointed and dejected, I decided I'd had enough of philosophy.

I spent the following year teaching English as a foreign language in Kraków in Poland. For the first few months, I studiously avoided philosophy, trying to lose myself in novels and popular science. But as time went on, I couldn't help myself and returned to reading philosophical articles on the topic of consciousness. It was during this time that I came across Thomas Nagel's classic 1972 article "Panpsychism"—a paper that never featured on my undergraduate reading list—and discovered that there was a neglected "third way" between materialism and dualism.

Panpsychism is the view that consciousness is a fundamental and ubiquitous feature of physical reality. This view is much misunderstood. Drawing on the literal meaning of the term—"pan" = everything, "psyche" = mind—it is commonly supposed that panpsychists believe that all kinds of inanimate objects have rich conscious lives: that your socks, for example, may currently be going through a troubling period of existential angst.

This way of understanding panpsychism is wrong on two counts. Firstly, panpsychists tend not to think that literally everything is conscious. They believe that the fundamental constituents of the physical world are conscious, but they need not believe that every random arrangement of conscious particles results in something that is conscious in its own right. Most panpsychists will deny that your socks are conscious, while asserting that they are ultimately composed of things that are conscious.

Secondly, and perhaps more importantly, panpsychists do not believe that consciousness *like ours* is everywhere. The complex thoughts and emotions enjoyed by human beings are the result of millions of years of evolution by natural selection, and it is clear that nothing of this kind is had by individual particles. If electrons have experience, then it is of some unimaginably simple form.

In human beings, consciousness is a sophisticated thing, involving subtle and complex emotions, thoughts, and sensory experiences. But there seems nothing incoherent with the idea that consciousness might exist in very simple forms.

We have good reason to think that the conscious experience of a horse is much less complex than that of a human being, and the experiences of a chicken less complex than those of a horse. As organisms become simpler perhaps at some point the light of consciousness suddenly switches off, with simpler organisms having no experience at all. But it is also possible that the light of consciousness never switches off entirely, but rather fades as organic complexity reduces, through flies, insects, plants, bacteria and amoeba. For the panpsychist, this fading-while-never-turning-off continuum further extends into inorganic matter, with fundamental physical entities—perhaps electrons and quarks—possessing extremely rudimentary forms of consciousness, to reflect their extremely simple nature.

Even given these qualifications, panpsychism still sounds crazy. It has a "New Age" feel that many just can't see past. But we should judge a view not by its cultural associations but by its explanatory power. Many widely accepted scientific theories are also crazily counter to common sense. Einstein's special theory of relativity tells us that time slows down at high speeds. According to standard interpretations of quantum mechanics, particles have determinate positions only when observed. And if Charles Darwin is to be believed, we have a common ancestor with apes! All of these views are wildly at odds with our commonsense view of the world, or at least they were when they were first proposed, but nobody thinks this is a good reason not to take them seriously. Why should we take common sense to be a good guide to how things really are?

But could such a view ever be tested? We can't look inside an electron and see its consciousness. And while there may be testable predictions associated with the view that inanimate objects have complex thoughts—for if they did, then we ought to be able to communicate with them—there don't seem any obvious behavioral implications of the thesis that electrons have extremely simple experiences. If it can be neither verified nor falsified, then panpsychism may seem to fall into the category of theories that the physicist Wolfgang Pauli rejected as "not even wrong."

In fact, there is some empirical support for panpsychism, which we'll get to later. But the main attraction of panpsychism is not its ability to account for the data of observation, but its ability to account for the reality of consciousness. We know that consciousness is real and so we have to account for it somehow. If a general theory of reality has no place for consciousness, then that theory cannot be true. What panpsychism offers us is a way of integrating consciousness into our scientific picture of the world, a way that avoids the deep problems associated with dualism on the one hand and materialism on the other.

Panpsychism avoids the problems of dualism because it does not postulate consciousness outside of the physical world and hence avoids the challenge of accounting for the interaction between the nonphysical mind and the physical brain. The panpsychist places human consciousness exactly where the materialist places it: in the brain. And because it is not trying to explain consciousness in terms of nonconscious brain processes, panpsychism also avoids the problems of materialism. Rather than trying to account for consciousness in terms of non-consciousness, the panpsychist aspires to explain the *complex* consciousness of human and animal brains in terms of *simple* forms of consciousness, simple forms of consciousness that are postulated to exist as fundamental aspects of matter.

Does this really count as an *explanation* of consciousness? Isn't this just taking consciousness for granted rather than genuinely explaining it? To be sure, panpsychism does not offer a *reductive* explanation of consciousness, that is to say, it does not explain consciousness in terms of something more fundamental than consciousness. However, it is a prejudice of materialism to suppose that this is obligatory. There is a great deal of precedent in science for nonreductive explanations which take the phenomenon to be accounted for as basic. Consider, for example, James Clerk Maxwell's theory of electromagnetism in the nineteenth century. Maxwell did not reductively explain electricity and magnetism in terms of the mechanistic properties and forces science was already committed to. Rather, he postulated new fundamental electromagnetic properties and forces and explained electromagnetism on that basis. Similarly, the panpsychist believes that the final theory of consciousness—when it comes along—will not explain consciousness in terms of something else but will rather take certain forms of consciousness as basic and build up from there.[1]

The reality of consciousness is a datum in its own right. If panpsychism offers the best explanation of that datum, then it is, to that extent, supported by the evidence. Hence, the central case for panpsychism rests on a form of inference to the best explanation, where the phenomenon to be explained is consciousness itself.

I cannot exaggerate the profound effect learning about panpsychism had on me. Here was a way of accepting the reality of consciousness—real, subjective, qualitative consciousness—which was entirely consistent with the facts of empirical science. Finally, I could resolve the intellectual tension between my scientific understanding and my self-understanding. In panpsychism I found intellectual peace; I could live comfortably in my own skin.

Moreover, I suddenly had a renewed enthusiasm for philosophy, and decided to take up graduate study the following September. At the time, not many philosophy departments in the U.K. had a panpsychist in residence. But the new professor at the University of Reading, Galen Strawson, was busy defending panpsychism as "the most parsimonious, plausible and indeed 'hard-nosed' position . . . about the nature of reality."[2] This seemed like the place for me. I had no idea at the time, but the writings of Strawson, and a little later my own writings, would eventually lead to a full-blown panpsychist renaissance in contemporary philosophy. At the start of my PhD, panpsychism was a position to be laughed at insofar as it was thought of at all. Fifteen years later, panpsychism has become a well-respected albeit minority position.

The roots of this revolution were the rediscovery, by Strawson and others, of crucial work on consciousness from the 1920s by the philosopher Bertrand Russell and the scientist Arthur Eddington. I am convinced that Russell and Eddington did for the science of consciousness what Darwin did for the science of life. It is to this topic we now turn.

SOMETHING LOST BETWEEN THE WARS

In May 1919 the English astronomer Arthur Eddington overturned more than two hundred years of scientific consensus by conducting the first experimental demonstration of Einstein's general theory of relativity. After the publication of the *special* theory of relativity in 1905, Einstein had spent ten years wrestling with the challenge of incorporating gravity into his revolutionary picture of the world; the *general* theory of relativity, published in November 1915, was the result of this decade of hard labor. In it, Einstein declared that Newton was wrong. Gravity was not, as Newton had supposed, a basic force. Instead gravity was to be explained in terms of a mutual causal interaction between matter and spacetime. Matter impacts on spacetime by altering its geometry, by curving it. The resulting curvature then impacts back on matter as material bodies have a tendency to follow the shortest path through spacetime, which is in turn determined by the curvature of spacetime. In other words, matter tells spacetime how to curve, while the curvature of spacetime tells matter how to move.

As Einstein published the general theory, Europe was more than a year into the Great War. As you might expect, the British at this time were not much interested in the speculative theories of German scientists, especially not ones that claimed to overturn the two-hundred-and-thirty-year hegemony of an English scientist. However, Eddington was a Quaker and an internationalist, and had little patience for nationalist prejudices. For Eddington, what mattered was the truth. During the First World War, he was secretary of the Royal Astronomical Society, and was granted exemption from military service on the basis of his scientific work being of national importance. As a result, Eddington was one of the first British scientists to absorb Einstein's complex new theory of gravity and was deeply enthused by what he read.

Only six months after the end of the war, together with the Astronomer Royal Frank Watson Dyson, Eddington conducted a series of observations of a solar eclipse from the island of Principe off the west coast of Africa. As the moon covered the sun, Eddington photographed stars visible around its covered edge. On the basis of this he was able to demonstrate that, precisely as Einstein's theory had predicted, the light from these stars had been bent by the spacetime curvature caused by the mass of the sun. Newton's theory did not predict that the starlight would be bent to this degree. Einstein was vindicated and overnight became an international celebrity.

Eddington went on to become a leading defender of relativity, not only persuading the scientific community of its truth but also reaching out to the public by finding vivid ways of explaining its meaning without the mathematical complexities. Einstein himself claimed that Eddington's writings on relativity were "the finest presentation of the subject in any language." When Eddington presented his proof of relativity at the Royal Society, it was joked that Eddington was one of the three men in the world who actually understood relativity. At first Eddington remained modestly silent, and upon being encouraged not to be so shy, he responded, "Oh no, I was wondering who the third one might be!"[3]

What is less remembered these days is that Eddington was a great supporter and popularizer not only of science but also of philosophy. This is less common among physicists these days, and indeed a number of prominent physicists have expressed hostility to philosophy. In their book *The Grand Design* Stephen Hawking and his co-writer, Leonard Mlodinow, began by declaring that "philosophy is dead" due to the fact that philosophers have failed to keep up with mathematical developments in modern physics. This declaration is somewhat ironic, as Hawking and Mlodinow go on in later chapters to indulge in philosophical discussions of free will and metaphysical realism. It is true that many philosophers—I would certainly put myself in this category!—wouldn't be able to handle the complex mathematical structures involved in general relativity. But this cuts both ways: most physicists have little grasp of the complex conceptual frameworks of contemporary philosophy.

The truth is that the skills that make one good at physics are not necessarily the skills that make one good at philosophy. What is required in philosophy is a certain capacity for thinking about everyday concepts—consciousness, justice, free will, knowledge—in abstraction from their everyday context. This is quite distinct from the mathematical skill essential for being a good physicist; in mathematics we abstract away entirely from these kinds of everyday concepts.

Eddington appreciated this, and understood that both the philosopher and the physicist have a role to play in uncovering the nature of reality. In his 1928 book, *The Nature of the Physical World,* Eddington speaks approvingly of the (at the time) cutting-edge philosophy of his contemporary Alfred North Whitehead, while expressing humility about his capacity as a nonspecialist to fully comprehend all of its nuances. In the course of this discussion, Eddington offers an intriguing metaphor to express what he saw as the common endeavor of the scientist and the philosopher:

> Although this book [Eddington's *The Nature of the Physical World*] may in most respects seem diametrically opposed to Dr. Whitehead's widely read philosophy of Nature, I think it would be truer to regard him as an ally who from the opposite side of a mountain is tunnelling to meet his less philosophically minded colleagues. The only thing is not to confuse the two entrances.[4]

What does all this have to do with consciousness? In science, Eddington was a relativity enthusiast. But in philosophy, Eddington was inspired by the novel solution to the problem of consciousness formulated in 1927 by the great English philosopher and Nobel Laureate Bertrand Russell. Russell had discovered a radical new approach that managed to circumnavigate the perennial debate between materialists and dualists. In the very same year that Russell published his view, Eddington used his Gifford lectures to expound his own take on it. These two men certainly fit Eddington's image of the scientist and the philosopher tunneling toward each other: as well as dealing with the problem of consciousness, Russell's 1927 book *The Analysis of Matter* offered philosophical reflections on the new scientific theories of relativity and quantum mechanics.

What is remarkable is that this exciting new approach to consciousness developed between the wars was almost completely forgotten about in the latter half of the twentieth century. After the Second World War, a materialist orthodoxy dominated science and philosophy for many years, to the extent that only a few mavericks dared to publish anything against it. The Russell-Eddington theory is not a form of dualism, but strictly speaking it isn't a materialist theory either, and so in these ideological times it was simply dismissed without argument. The view we are about to discuss was certainly not something on my syllabus when I was a philosophy undergraduate.

Things are slowly changing, and there is now a healthy and respected minority of academic philosophers defending "antimaterialist" views of one form of another. According to a recent survey of academic philosophers, materialists are still in the majority but only just (56.5 percent).[5] This gradual shift has led to a rediscovery of the Russell-Eddington middle way between materialism and dualism, and in academic philosophy this is now widely seen as one of the most promising ways forward on the problem of consciousness. But the view that has emerged, which can be understood as a form of panpsychism, is still almost completely unknown outside of the ivory tower of academic philosophy. Without the input of empirical scientists, the view is at this stage somewhat schematic and underdeveloped. One main aim of this book is to convey to a broader audience the plausibility of the Russell-Eddington framework for explaining consciousness, with the hope that as a community of co-investigators we might begin to fill in the details.

WHY PHYSICS TELLS YOU LESS THAN YOU THINK

In the public mind, physics is on its way to giving us a complete account of the nature of space, time, and matter. We are not there yet of course; for one thing, our best theory of the very big—general relativity—is inconsistent with our best theory of the very small—quantum mechanics. But it is standardly assumed that one day these challenges will be overcome and physicists will proudly present an eager public with the Grand Unified Theory of everything: a complete story of the fundamental nature of the universe.

This is not how Russell and Eddington saw physics. To the surprise of the scientific community, they argued that physics had been so successful precisely because it had stopped trying to tell us anything about the nature of matter. For ease of exposition, I will focus on Eddington's presentation of this argument, which begins with a homely example:

> If we search the examination papers in physics and natural philosophy for the more intelligible questions we may come across one beginning something like this: "An elephant slides down a grassy hill-side. . . ." The experienced candidate knows that he need not pay much attention to this; it is only put in to give an impression of realism. He reads on: "The mass of the elephant is two tons." Now we are getting down to business; the elephant

> fades out of the problem and a mass of two tons takes its place. What is this two tons, the real subject-matter of the problem? . . . [it is] the reading of the pointer when the elephant was placed upon a weighing machine. Let us pass on. "The slope of the hill is 60°." Now the hill-side fades out of the problem and an angle of 60° takes its place. What is 60°? There is no need to struggle with mystical conceptions of direction; 60° is the rendering of a plumb line against the divisions of a protractor. . . . And so we see that the poetry fades out of the problem, and by the time the serious application of exact science begins we are left with only pointer readings.[6]

In 1623 Galileo declared that mathematics was to be the language of science. In the above quotation of 1928 we find Eddington fully appreciating, perhaps for the first time in the history of modern science, what this amounts to.

To try to make Eddington's point clear, let us compare the equations of physics to the equations of economics. Consider the following very simple equation from economic theory:

$$A=T/L$$

In the above equation, A stands for average product, T for total product, and L for labor. So the equation tells us that the average product is equal to the total product divided by labor input. To take a concrete example, if a factory is producing 100 widgets a day (the total product) using 10 workers (the labor input), then the average product is 10.

Note that the equation doesn't tell us what "labor" or "product" are. Rather it relies on our having a pre-theoretical understanding of these notions in order to assert a mathematical relationship between them. If an alien came across an economics textbook but had no understanding of what labor and product are, then this equation would be meaningless to it.

Something similar is going on in the equations of physics. Consider Newton's law of universal gravitation:

$$F = G\frac{m_1 m_2}{r^2}$$

The variables m_1 and m_2 stand for the masses of two objects that we want to work out the gravitational attraction between, F is the force between those two masses, G is the gravitational constant (a number we know from observation), and r is the distance between m_1 and m_2. Just as the economics equation didn't tell us what "labor" and "product" are, so the above physics equation doesn't tell us what "mass," "distance," and "force" are. This is not something peculiar to Newton's law.* The subject matter of physics is the basic properties of the physical world: mass, charge, spin, distance, force. But the equations of physics do not explain what these properties are. They simply *name* them in order to tell us about the relationships that obtain between them.

It is not a problem that the equations of economics do not tell us what, for example, "labor" or "price" are, as economics is not a purely mathematical science. It presupposes nonmathematical definitions of its central concepts. We could define "labor," for example, as work done to produce goods and services, where we rely on a nonmathematical understanding of what work is and of what goods and services are. In contrast, physics since Galileo has been a purely mathematical

* One might be tempted to think that Einstein's theory of gravity avoids these difficulties. I argue in Technical Appendix B that it doesn't.

science. There is nothing beyond the equations, and hence no resources with which to define what "mass," "charge," etc., are. Mathematical physics simply does not have the resources to tell us what the basic features of the physical world are.

If physics is not telling us the nature of physical reality, what *is* it telling us? The crucial point Eddington is trying to convey with his talk of "pointer readings" is that *physics is a tool for prediction.* Even if we don't know what "mass" and "force" really are, we are able to recognize them in the world. They show up as readings on our instruments, or otherwise impact on our senses. And by using the equations of physics, such as Newton's law of gravity, we can predict what's going to happen with great precision (although strictly speaking Newton's law has been displaced by general relativity, it's still generally used as it's accurate enough for most purposes). It is this predictive capacity that has enabled us to manipulate the natural world in extraordinary ways, leading to the technological revolution that has transformed our planet.

We can put this loosely (although see Technical Appendix B for a more careful analysis) by saying that physics tells us not what matter *is* but only what it *does.* Think about an electron. What does physics tell us about an electron? Physics tells us that an electron has mass and negative charge (among other properties). How does physics characterize mass and negative charge? Mass is characterized in terms of the disposition to attract other things with mass (that's what we call "gravity") and in terms of the disposition to resist acceleration (the more mass something has the harder it is to get it to move, to stop moving, or to change velocity). Negative charge is characterized in terms of the disposition to repel other negatively charged things and to attract positively charged things. Note that all of this concerns the behavior of the electron, what it does in relation to other physical particles. And the same is true for everything else physics tells us about the electron. Physics confines itself to telling us only what an electron does.

And yet intuitively there must be more to what an electron is than what it does. An electron must have an *intrinsic nature,* as philosophers like to say. Consider the following analogy. Imagine you have a chess piece on a board. You may know what the chess piece does, how it behaves; if it's a bishop, let's say, it moves diagonally. But there must be more to the nature of the chess piece than what it does. There must be some way the chess piece is in and of itself, independent of its behavior. It may, for example, be made of wood or plastic. When we ask how the chess piece is in and of itself, we are asking about its intrinsic nature. Similarly for the electron, independently of what it does in relation to other particles, there must be some way the electron is in and of itself. And yet physics leaves us completely in the dark about the intrinsic nature of the electron.

This is referred to as "the problem of intrinsic natures":

> *The Problem of Intrinsic Natures*—Physical science restricts itself to providing information about the behavior of the things it talks about—particles, fields, spacetime—and tells us nothing about their intrinsic natures.

(Readers skeptical of the need for intrinsic natures might want to read Technical Appendix B, at the end of this chapter, before continuing.)

The problem of intrinsic natures arises not just for fundamental physics, but also in the "higher-level" sciences of chemistry and neuroscience. In neuroscience, physical processes in the brain are characterized either in terms of their *causal role* in the brain (i.e., what they do in relation to other parts of the brain and in relation to behavior) or in terms of their chemical constitution (e.g., in terms of neurotransmitters, such as amino acids, peptides, and monoamines). In chemistry, elements and molecules are characterized either in terms of their *causal relationships* with other chemical entities (e.g., acids are defined in terms of their capacity to donate protons or hydrogen ions and/or accept electrons) or in terms of their *physical constituents* (e.g., water is made up of molecules with two hydrogen atoms and one oxygen atom). And so ultimately, from the starting points of neuroscience and chemistry, we get to physics, where, as we have already discussed, basic physical properties are characterized in terms of what they do. All the way up the hierarchy of the physical sciences we learn only about causal relationships: about what physical things do.

This takes a while to absorb. We are used to thinking that physical science is telling us about the "stuff" of the world. When we learn that water is H_2O, or that heat is molecular motion, we are inclined to think that we are discovering the real nature of water and heat. This is partly because chemistry characterizes these phenomena by identifying their atomic components, and so we feel we are learning something when we come to know that water is made up of hydrogen and oxygen. It is only when we get down to the physical explanation of what "hydrogen" or "oxygen" are that we discover (i) that chemistry characterizes hydrogen and oxygen entirely in terms of their physical components, but also (ii) that physics tells us absolutely nothing about the intrinsic nature of those components. There is a very real sense in which we have no idea what hydrogen and oxygen are, and hence we have no idea what water is!

We began this section with the public perception that physics is giving a complete account of the universe around us. We can now see how wrong that public perception is. Far from being complete, there is a gaping hole in the theory of the universe we get from physical science. Even if physicists one day manage to unify general relativity and quantum mechanics and present us with the Grand Unified Theory, this theory would still not be complete. For it would tell us nothing about the intrinsic nature of anything that exists.

In spite of his later philosophobia, Stephen Hawking appreciated this point in *A Brief History of Time,* the book which made him a household name:

> Even if there is only one possible unified theory, it is just a set of rules and equations. What is it that breathes fire into the equations and makes a universe for them to describe?[7]

The equations of physics allow us to predict the behavior of matter with great precision. But it is the intrinsic nature of matter that breathes fire into those equations. And on this topic physics has nothing to say.

TOWARD A BROADER CONCEPTION OF SCIENCE

It is important to emphasize that Russell and Eddington were not making a criticism of physics itself. It's not a fault of physical science that it doesn't postulate intrinsic natures. That's not what it's supposed to do. The purpose of physical science is to predict behavior and it does that perfectly well. Philosophers who raise the problem of intrinsic natures are not trying to tell physical scientists that they need to do their job differently.

What is being exposed here is a certain popular perception of physics, according to which the purpose of physics is to hold up a mirror to reality. This is simply not what physical science is in the business of.

In recent years we have gotten used to taking "science" and "physical science" to be synonymous. At the same time, we look to scientists to give us a complete theory of reality. These two demands on science cannot be reconciled. So long as "science" is equated with "physical science," it will be subject to the following limitations:

- It will be unable to account for consciousness, as the qualitative reality of consciousness cannot be captured in the quantitative language of physical science.
- It will be confined to telling us what matter does, remaining silent on its intrinsic nature.

The solution is to move to a more expansive conception of science, one that incorporates physical science as an aspect. The quantitative sciences of physics and chemistry have had great success, but this is in part because they were designed to fulfill a specific and limited goal: the prediction of behavior.

A more expansive science may not be more useful, at least not in the sense of helping us build bridges or cure cancer. Physical science is the really useful bit of science because it provides us with detailed information about how things behave. But we shouldn't confuse that practical utility with the ontological aspiration of giving a complete theory of reality. Without going beyond the information provided by physical science we will be unable to achieve the ultimate goal of science: a Theory of Everything.

What would this new science look like, and how would it do a better job of accounting for consciousness? We will get to this. But first we need to understand how Russell and Eddington solved the problem of consciousness.

TURNING THE PROBLEM OF CONSCIOUSNESS UPSIDE DOWN

Physics tells us nothing about the intrinsic nature of matter. Are we then forced to conclude that we know nothing of the intrinsic nature of matter? Not quite, according to Eddington:

> We have dismissed all preconception as to the background of the pointer readings [by which Eddington means the causes of readings on our measuring instruments], and for the most part we can discover nothing as to its nature. But in one case—namely, for the pointer readings of my own brain—I have an insight which is not limited to the evidence of the pointer readings. That insight shows that they are attached to a background of consciousness.[8]

In other words, I have but one small window onto the intrinsic nature of matter: I know that the intrinsic nature of the matter inside my brain involves consciousness. I know this because I am directly aware of the reality of my own consciousness. And, assuming dualism is false, this reality I am directly aware of is at least part of the intrinsic nature of my brain.*

* As my former teacher Galen Strawson is always keen to emphasize, one difficulty here is that we very easily slip back into the dualist ways of thinking, whereby we implicitly suppose that our experience is *produced* by the physical properties of the brain rather than being literally identical with

This realization turns the problem of consciousness on its head. It is generally assumed that neuroscience gives us significant understanding of the nature of the brain, and that the challenge is to understand how the mysterious phenomenon of consciousness "fits in" to the more comprehensible reality revealed by physical science. In fact, far from being a mystery, consciousness is the only bit of physical reality we really understand. It is the rest of the physical world that is a mystery. As Eddington put it:

> We are acquainted with an external world because its fibres run into our own consciousness; it is only our own fibres that we actually know; from these ends we more or less successfully reconstruct the rest, as a palaeontologist reconstructs an extinct monster from its footprint.[9]

This points the way to an elegant solution to the problem of consciousness. In essence, the problem of consciousness can be put as follows:

> *The Problem of Consciousness*—How do we integrate consciousness into our scientific story of the universe?

---

them. Suppose you are currently smelling fish; if dualism is false, then that fishy experience is part of the intrinsic nature of your living brain at that moment. In *The Analysis of Matter*, Russell put the point vividly as follows: "What the physiologist sees when he examines a brain is in the physiologist, not in the brain he is examining. What is in the brain by the time the physiologist examines it if it is dead, I do not profess to know; but while its owner was alive, part, at least, of the contents of his brain consisted of his percepts, thoughts, and feelings" (p. 320). We can also see from this quotation that Russell does not fully embrace the panpsychist position, as we will discuss below.

In this chapter we have encountered another problem, one that on the face of it has nothing to do with consciousness:

> *The Problem of Intrinsic Natures*—Physics tells us nothing of the intrinsic nature of matter.

The brilliant insight of Eddington, building on Russell, was to solve both of these problems at once:

> Problem 1: We need a place for consciousness.
> Problem 2: We have a huge hole at the center of our scientific story.
> Solution: Plug the hole with consciousness.

In other words, Eddington's proposal is that consciousness is the intrinsic nature of matter. It is consciousness, for Eddington, that breathes fire into the equations of physics.

Here is the idea. Physics characterizes mass and charge "from the outside" (in terms of what they do) but "from the inside" (in terms of their intrinsic nature) mass and charge are incredibly simple forms of consciousness. Moving up a level, chemistry characterizes chemical properties "from the outside," but "from the inside" they are complex forms of consciousness derived from the basic forms of consciousness found at the level of fundamental physics. Moving up another level, neuroscience characterizes brain processes "from the outside," but "from the inside" they are states of human experience, incredibly complicated forms of consciousness derived from the more basic forms of consciousness found at the levels of chemistry and physics.

This seems a coherent proposal. But what reason do we have to take it seriously? The first thing to say is that it's not clear that there's an alternative. This is because, strange as it might sound at first hearing, there doesn't seem to be a candidate for being the intrinsic nature of matter other than consciousness. Certainly, physical science does not furnish us with an alternative option, given its silence on the intrinsic nature of matter (see Technical Appendix B for more detail). And if neither introspection nor observation gives us a clue as to the intrinsic nature of matter, where else are we supposed to look? The choice seems to be between the panpsychist view as to the intrinsic nature of matter, and the view that matter is, as the seventeenth-century philosopher John Locke put it, "we know not what." Insofar as we seek a picture of reality without gaps, panpsychism may be our only option. For Eddington, this was enough to embrace panpsychism:

> The Victorian physicist felt that he knew just what he was talking about when he used such terms as *matter* and *atoms*. Atoms were tiny billiard balls, a crisp statement that was supposed to tell you all about their nature in a way which could never be achieved for transcendental things like consciousness, beauty, or humour. But now we realise that science has nothing to say about the intrinsic nature of the atom. The physical atom is, like everything else in physics, a schedule of pointer readings. The schedule is, we agree, attached to some unknown background. Why not then attach it to something of a spiritual [i.e., mental] nature of which a prominent characteristic is thought [by which Eddington means consciousness]. It seems rather silly to prefer to attach it to something of a so-called "concrete" nature inconsistent with thought, and then to wonder where thought comes from.[10]

Furthermore, there is a strong case that panpsychism is the simplest theory consistent with what we directly know about the nature of matter. Eddington's starting point is as follows:

1. Physical science tells us absolutely nothing about the intrinsic nature of matter, and
2. The only thing we know about the intrinsic nature of matter is that some of it, i.e., the matter inside brains, has an intrinsic nature made up of forms of consciousness.

It is hard to really absorb these two facts, as they are diametrically opposed to the way our culture thinks about science. But if we manage to do so, it becomes apparent that the simplest hypothesis concerning the intrinsic nature of matter *outside of brains* is that it is continuous with the intrinsic nature of matter *inside of brains,* in the sense that both inside and outside of brains matter has an intrinsic nature made up of forms of consciousness. To deny panpsychism one would need a reason for supposing that matter has two kinds of intrinsic nature rather than just one.

I call this the "simplicity argument" for panpsychism. This might seem like an insubstantial consideration, but in fact considerations of simplicity play an important role in science. As we discussed in chapter 2, there are always an infinite number of theories consistent with the evidence, and we must choose between them on the basis of simplicity: don't believe in more things if you can get away with fewer. This is, after all, the main reason that most scientists and philosophers reject the existence of immaterial minds. And, as we also discussed in chapter 2, it is on the basis of considerations of simplicity that the scientific community almost universally prefers Einstein's theory of relativity over the Lorentzian alternative.

Indeed, there is a further analogy between panpsychism and special relativity. While special relativity is simpler and more elegant than its Lorentzian rival, it is also much more contrary to common sense, as it entails all sorts of peculiar things about the nature of time, for example, that time passes slower at higher speeds. Despite the inelegance of the Lorentzian view, it allows us to preserve our commonsense view of time, which Lorentz himself felt unable to give up on. It is good scientific practice to go with the theory that is simple and elegant rather than the theory that preserves common sense. Those who follow this dictum with rigorous objectivity and with a mind free from bias will be led, I believe, to panpsychism.

While in the mind-set that physics is on its way to giving us a complete picture of the nature of space, time, and matter, panpsychism is absurd, as physics does not attribute experience to fundamental particles. But once one absorbs the problem of intrinsic natures, the universe looks very different. All we get from physics is this big black-and-white abstract structure, which we must somehow fill in with intrinsic nature. We know how to color in one bit of it: the brains of living organisms are colored in with consciousness. How to color in the rest? The most elegant, simple, sensible option is to color in the rest of reality with the same pen.

It is important to emphasize that what we are considering is a deeply *nondualistic* form of panpsychism. When we first think about panpsychism, we tend to conceive of it in dualist terms, as though the electron has its physical properties—mass, charge, spin, etc.—*and* its consciousness properties sitting side by side, as it were. Such a dualistic form of panpsychism would share many of the problems of conventional dualism that were discussed in chapter 2. There would be a cost in terms of simplicity, as we would be adding nonphysical properties to matter in addition to its physical properties. And more importantly, neuroscience shows no sign of the causal effects of mysterious nonphysical properties in the brain, which arguably gives us strong grounds for thinking there aren't any.

Eddington's panpsychism was not dualistic. His view was not that particles have two sets of properties: physical properties (mass, charge, spin, etc.) on the one hand and nonphysical consciousness properties on the other. The view is rather that the physical properties of a particle (mass, spin, charge, etc.) *are themselves* forms of consciousness. Those very properties that physical science characterizes behavioristically are, in their intrinsic nature, forms of consciousness. Physics characterizes mass "from the outside," in terms of what it does, but in its intrinsic nature mass is a form of consciousness. At the very least, this view is no less simple than the view that mass has some entirely unknown nature, which is arguably the only alternative to panpsychism once we accept that physics is silent on the intrinsic nature of matter.* Thus, Eddington's panpsychism does not *add* to our theory of matter; it merely offers a positive proposal as to what matter essentially is.

More generally, Eddingtonian panpsychism gives us an

* Sam Coleman offers a nonpanpsychist proposal regarding the intrinsic nature of matter, according to which it is constituted of *unexperienced qualities* (see, for example, his article "Panpsychism and Neutral Monism"). I have argued against this proposal in my academic book, *Consciousness and Fundamental Reality*. Another alternative is to deny the need to postulate any kind of intrinsic nature to matter; see Technical Appendix B for my response to this position.

elegant way to unify mind and matter, and thus to avoid altogether the irresolvable dispute between dualism and materialism. The dualist offers us a radically disunified picture of reality, with no obvious way of explaining how mind and brain interact. Materialism brings unity, but at the cost of having no place for consciousness. Eddington's panpsychism avoids all of these problems. It has all the simplicity and unity of materialism, but it also has a place for consciousness.

The basic idea of explaining consciousness in terms of the intrinsic nature of matter came from Russell. However, Russell's own way of spelling this out was not quite full-blown panpsychism. Russell believed that the intrinsic nature of the world is constituted of a third element, neither mental nor physical but more akin to the former than the latter. This view is known as "neutral monism" and the resurgence of interest in the Russell-Eddington approach has involved defenses both of Russell's neutral monist version and of Eddington's panpsychist version.*
In my view, although both are worth exploring, considerations of simplicity favor the latter.

The focus of the first chapter was how Galileo cast out the sensory qualities from the physical world in order to allow for the possibility of mathematical physics. We can now appreciate what Galileo's error really was. He thought that mathematics could provide insight into the nature of physical reality, and that the nature it revealed was incompatible with the reality of the sensory qualities (which must therefore exist in the soul). In fact, he was wrong on both counts: mathematical models don't tell us anything about the intrinsic nature of matter, but

* Some prominent contemporary defenders of neutral monism are Sam Coleman, Tom McClelland, Susan Schneider, and Daniel Stoljar.

for precisely that reason nor do they exclude the reality of the sensory qualities. In 1623 Galileo took the sensory qualities out of the physical world. Three hundred years later in 1927 Russell and Eddington finally found a way of putting them back.

A CULTURE SHIFT

As I have already mentioned, when I began my PhD, panpsychism was not really given the time of day. Even five years later when I began to apply for academic jobs, I was told to keep my panpsychist views secret in case they reduced my chances of employment. However, intellectual fashions change and the last five or ten years have seen the science of consciousness become a lot more open to panpsychism and closely related views. In this section, we will consider a few examples.

In chapter 2 we discussed the Integrated Information Theory of consciousness—or IIT for short—a leading neuroscientific theory of consciousness according to which consciousness is associated with the integrated information in a physical system. What I did not mention there was that IIT has panpsychist implications, implications which its creator, Giulio Tononi, freely accepts.

Almost any physical system is associated with *some* degree of integrated information, even a single molecule. The presence of integrated information does not in itself indicate the presence of consciousness, according to IIT. The theory tells us that, in any physical system, consciousness is present at the level at which there is the most integrated information. For example, a molecule contained within the brain will not be conscious, as the level of integrated information in a brain is much higher than the level of integrated information in a single molecule.

However, IIT predicts that a molecule in a puddle of water will be conscious, as there is more integrated information in the molecule than in the puddle as a whole. This is a theory of consciousness with considerable empirical support, and yet it entails that consciousness is much more widespread than we ordinarily assume.

The psychologist Susan Blackmore is a renowned skeptic. As a young woman in the 1960s she had a vivid out-of-body experience which persuaded her of the reality of paranormal phenomena, and she went on to complete a PhD in parapsychology defending this belief. However, the more she investigated psychic phenomena, the more she became persuaded that belief in such things resulted from "wishful thinking, self-deception, experimental error and, occasionally, fraud."[11] Her later work focused on memes and evolutionary theory rather than telepathy and telekinesis. While Blackmore is still committed to a fiercely empirical and nonsupernaturalist approach to the mind, she now takes panpsychism to be a serious option: "So long as a physical system can differentiate itself from its environment," she told me at a philosophy event in Hay-on-Wye in Wales in 2018, "that system could be said to have experience."

One of the philosophers I've clashed most with over the last fifteen years is David Papineau. A professor at King's College London, David is one of the most well-known critics of the arguments I defended in the last chapter—involving zombies and Black and White Mary—which try to demonstrate that consciousness cannot be explained by physical science. Despite our deep disagreements, I have learned a lot from our conversations, which have taken place informally as well as in philosophy podcasts, and we have become close friends as well as intellectual opponents. I very much admire David's ability to separate intellectual disagreement from personal antipathy, a distinction too often overlooked in contemporary public discourse.

However, I was flabbergasted—and I don't use that word lightly—when we last debated to hear David say that he was now drawn to embrace a form of panpsychism![12] His motivations are very different compared to my own. I'm motivated by the failure of physical science to account for consciousness, and consequently by the need to find some other way to fit consciousness into our scientific picture of the world. David denies there is any problem here: one's consciousness is just a matter of the chemical processes in one's brain and that's all there is to it. Instead, David is drawn to panpsychism precisely because he thinks there is *nothing special* about consciousness:

> . . . if consciousness were constituted by some extra mind stuff, something additional to the physical realm, then there would be a real difference between the presence and absence of this mind stuff. However, once we free ourselves from the intuitive myth of such extra mind stuff, should we continue to think of consciousness as constituting a distinctive physical kind?[13]

In other words, those who think of consciousness as only in the brain are implicitly thinking in dualist terms, as though consciousness were some special magical substance that only comes about in very special circumstances. Consciousness scientists, for Papineau, are like the alchemists of old looking for the unique conditions that could transform copper into gold. Once we've completely rid ourselves of the idea that there is something special about consciousness, there seems no reason not to identify it with some utterly mundane and commonplace physical process. Consciousness might be nothing more than mass or electrical charge, physical properties that are ubiquitous in the physical world.

On this basis, Papineau is led to believe that consciousness probably exists wherever there are physical processes occurring. But don't we have very good scientific grounds for thinking that consciousness disappears in, say, comas or dreamless sleep? Certainly there are periods of deep sleep from which consciousness, if there is any, cannot be *remembered*. But the fact that no consciousness was remembered does not in itself show that no consciousness was present.

Waking experience is preserved by memory and capable of being recalled throughout the day. At each moment, I am aware of what I experienced the previous moment. And asked what I was experiencing earlier in the day, I can easily access that information, at least in rough outline. In this way, one's waking experience is tightly bound together by memory. If your experience suddenly changed, for example, if you suddenly found yourself on top of a mountain in fierce weather, you'd immediately notice the difference between the experience you are currently having and the experience you had a moment earlier.

Even in the dreams we do remember when we wake up, what is experienced from moment to moment is often not so tightly bound together by memory. One moment we're back in high school being taught French by Miss Clarke, and the next moment we're on top of a mountain without noticing anything has changed. Memory is still recording the dream (if it weren't we wouldn't be able to remember it upon waking), but it is not binding moment-to-moment experience into a coherent whole as it does in waking life.

Now, it could be that in periods of sleep we call "dreamless" there is still experience, but experience which memory takes absolutely no record of. What is experienced in one moment is forgotten in the next. There could be no narrative in such a state, just fleeting images and shapes, appearing and vanishing in quick succession. This would allow Papineau to maintain his view that consciousness is present everywhere, even in what we think of as dreamless sleep.

One might understandably be suspicious of any move from "we don't know whether there's consciousness in deep sleep" to "there is consciousness in deep sleep." I don't know for certain that there isn't a gremlin living in the darkness of my bathroom during the night that disappears every time the door is opened or the light is turned on, but in the absence of any positive reason to believe the gremlin does exist I quite sensibly suppose it doesn't. This is Ockham's razor (discussed in chapter 2) in action, the principle that we should believe in the fewest number of entities consistent with our evidence.

But here's Papineau's crucial point: Ockham's razor tells us not to believe in *extra* entities that are surplus to requirements. If we are tempted to apply this principle to "dreamless" sleep to rule out the existence of experience, this shows that we're taking experience to be something *extra* to the physical brain processes already going on during these periods. But this is precisely what Papineau is denying. Experiences *just are* physical brain processes. Given this, the theoretical drive for simplicity now points in the opposite direction: it is simpler to suppose that brain processes continue to be experiences when we fall into deep sleep than to suppose that they change their nature and cease to be experiences. This is something like the "simplicity argument" for panpsychism I defended above, and it is fascinating that Papineau has reached the same conclusion from a materialist point of view.

Papineau is also making a more general point about the science of consciousness. A lot of energy is spent in neuroscience trying to track the "neural correlates of consciousness," i.e., the brain processes that correlate with conscious experience. Papineau's view is that consciousness scientists are inadvertently tracking the wrong property. They are not really homing in on the neural correlates of *consciousness* but on the neural correlates of *remembered consciousness*, i.e., states of experience of which memory keeps a record. He puts the point as follows:

> It is tempting to think of our introspective gaze as being attracted by some kind of inner illumination. The reason some states, but not others, are accessible, we suppose, is because they glow with a special light. But this isn't the only way to see things. By way of analogy, consider the items that appear on the television news. We don't think that they are distinguished from the ordinary run of events by some distinctive radiance. They are just events that happen to attract the attention of the cameras. Similarly there is no reason to think of our conscious states as being distinguished by some extra lustre. They appear to us as they do in virtue of our having access to them, not because they have some distinctive luminosity.[14]

Papineau is not trying to explain the fact that a state is conscious in terms of the fact that we can access it through introspection and memory. His claim is that consciousness is not some special "glowing" physical property but is rather identical with some utterly mundane and ubiquitous physical property, perhaps mass or electric charge. Some conscious states are rendered accessible by memory and introspection; others are not. But this difference does not mark out a fundamental or significant divide in nature. If we really want to deny that consciousness is special and magical, then we should get rid of the idea that very special circumstances are required for its existence.

MAKING UP MINDS

By now I'm sure you're all convinced that panpsychism is a wonderful theory that single-handedly solves the problem of consciousness once and for all. I'm sorry to say that there are difficulties for the view I am defending. Most renowned is the challenge that has become known as the "combination problem." The combination problem is posed by the following question: How do you get from little conscious things, like fundamental particles, to big conscious things, like human brains? We understand how bricks make up a wall, or mechanical parts make up a functioning car engine. But we are at a loss to understand how little minds could somehow combine to make up a big mind.

This problem for panpsychism was first articulated by the nineteenth-century* psychologist William James, who referred to panpsychism as "the mind dust theory":

> Take a hundred . . . [feelings] . . . shuffle them and pack them as close together as you can (whatever that may mean); still each remains the same feeling it always was,

* Although the recent interest in panpsychism is prompted by the work of Russell and Eddington, the view has a rich history and enjoyed something of a heyday in the nineteenth century. For more on the history of panpsychism in Western philosophy, see David Skrbina's *Panpsychism in the West* and the "Panpsychism" entry of *The Stanford Encyclopedia of Philosophy* by Goff et al. (the latter is available online).

> shut in its own skin, windowless, ignorant of what the other feelings are and mean. There would be a hundred-and-first feeling there, if, when a group or series of such feelings were set up, a consciousness belonging to the group as such should emerge. And this 101st feeling would be a totally new fact; the 100 feelings might, by a curious physical law, be a signal for its creation, when they came together; but they would have no substantial identity with it, nor it with them, and one could never deduce the one from the others, nor (in any intelligible sense) say that they evolved it.[15]

If you take a bunch of Lego bricks and arrange them in the right way, you've thereby got a Lego tower. Nothing more needs to be added for the parts to make the whole. But if you stick 100 tiny conscious minds together, it's hard to see how you'd have anything more than 100 tiny conscious minds, each existing in splendid isolation from each of the others. How could we make sense of the 100 minds *blending* or *fusing* into a unified whole? James continues with his vivid analogies:

> Take a sentence of a dozen words, and take twelve men and tell to each one word. Then stand the men in a row or jam them in a bunch, and let each think of his word as intently as he will; nowhere will there be a consciousness of the whole sentence. We talk of the "spirit of the age," and the "sentiment of the people," and in various ways we hypostatize "public opinion." But we know this to be symbolic speech, and never dream that the spirit, opinion, sentiment, etc., constitute a consciousness other than, and additional to, that of the several individuals whom the words "age," "people," or "public" denote.

> The private minds do not agglomerate into a higher compound mind.[16]

Solving the combination problem is crucial. After all, what we ultimately want explained is *human consciousness,* or more generally the consciousness of human and nonhuman animals. This kind of consciousness is our theoretical starting point, and the success or failure of consciousness science is judged in terms of how well or how badly it can account for this. The materialist tries to account for animal consciousness in terms of purely physical states of the brain. The panpsychist tries to account for animal consciousness in terms of particle consciousness. But if the latter is no more successful than the former, then panpsychism is a lost cause.

While this is certainly a deep problem, it would be wrong to think of it as in the same ballpark as the problems facing materialism. The challenge for the materialist is to bridge the gap between the *objective quantities* of physical science and the *subjective qualities* of conscious experience. But as we saw in the last chapter, this project is of dubious coherence, and it is certainly not something we have made the slightest progress on.

The combination problem, in contrast, is more tractable; it is simply the challenge of getting from *simple* subjective qualities to *complex* subjective qualities. Most panpsychists would accept that we do not yet have an entirely satisfactory solution to this problem, but, in contrast to the case of materialism, there are no worries about the *logical coherence* of the project.*

* Some people argue that the combination problem is particularly challenging when we focus not on consciousness but on the things that have consciousness. Maybe we can make sense of simple forms of consciousness combining to form complex forms of consciousness, but what is especially hard to make sense of (on this view) is lots of conscious *minds* combining

While both the materialist and the panpsychist have a gap to close, the panpsychist—in contrast to the materialist—is trying to bridge a gap between two things of essentially the same kind:

- *The Materialist Gap*—Between the objective quantities of physical science and the subjective qualities of consciousness.
- *The Panpsychist Gap*—Between the simple subjective qualities postulated to exist at the micro-level and the complex subjective qualities we know to exist in human and animal brains.

Some people dismiss panpsychism simply because the combination problem has not yet been solved. To my mind this is like someone in 1859 rejecting Darwinism on the basis that *On the Origin of Species* did not contain a completely worked out history of the evolution of the human eye. Darwin's theory of evolution by natural selection was a broad theoretical

---

to form an überconscious mind. Indeed, Sam Coleman, in his article "The Real Combination Problem," has argued that we cannot make coherent sense of minds combining. However, there is something strange about the separation of these two ways of thinking about the combination problem. You can't have a feeling just floating around without a conscious mind to experience it, just as you can't have a shape floating around without an object whose shape it is. The existence of an experience trivially entails the existence of an experiencer (just as the existence of a shape trivially entails the existence of a shaped object). So if we can account for the presence of complex, macro-level forms of consciousness (in terms of facts about micro-level consciousness), we have thereby accounted for the presence of complex, macro-level conscious minds (in terms of facts about micro-level consciousness). The theories we will examine below seem to me to have the potential to do the former, and we can infer on that basis that they also have the potential to do the latter. There is much more detail on the combination problem, including a response to Coleman's argument, in my academic book, *Consciousness and Fundamental Reality.*

framework, and it has taken decades to fill in the details and we still have a long way to go. Likewise, it will take many decades of interdisciplinary collaboration to fill in the details of the broad theoretical framework for explaining consciousness devised by Russell and Eddington. And while a vast amount of time and money has been plowed into trying to find a materialist solution to the problem of consciousness, the project of trying to find a solution to the combination problem has only just begun.

Attempting to solve the combination problem is currently the main focus of the panpsychist research program, and already there are a number of promising proposals. In what follows I will give a flavor of two of the leading avenues of research.

THE DIVIDED BRAIN

The conscious life of a person is deeply unified. There are many different aspects of my current experience—the visual experience of the laptop in front of me, the smell of coffee, the sound of a couple arguing in the next apartment—but they are all had by a single subject of experience that I call "me." My experience of the coffee smell doesn't happen in an isolated "pocket" of the mind separate from the sound of the couple arguing. Rather I have a single unified experience involving each of these aspects. Or at least, that's how things seem.

This deep unity to the mind was one of the things that convinced Descartes that the mind could not possibly be a material entity. While material things like bodies can always be divided, according to Descartes, the idea of dividing up a mind makes no sense at all:

> . . . the body is by its very nature always divisible, whilst the mind is utterly indivisible. For when I consider the mind, or myself in so far as I am merely a thinking thing, I am unable to distinguish any parts within myself; I understand myself to be something quite single and complete. Although the whole mind seems to be united to the whole body, I recognise that if a foot or arm or any other part of the body is cut off, nothing has thereby been taken away from the mind. As for the faculties of willing, of understanding, or sensory perception and so on, these cannot be termed parts of the mind, since it is one and the same mind that wills, and understands and has sensory perceptions. . . . This one argument would be enough to show me that the mind is completely different from the body, even if I did not already know as much from other considerations.[17]

What Descartes is describing certainly reflects how things seem to be "from the inside." There seems to be just one "me" that sees and hears, thinks, and speaks on the basis of its thoughts. However, pioneering experiments in the twentieth century by Roger Sperry, for which he eventually received the Nobel Prize in 1981, revealed that the core functions we associate with our sense of self are performed in very different regions of the brain.[18] For example, the speech control center is located in the left half of the brain, while the ability to recognize faces in located in the right half of the brain. Moreover, Sperry discovered this through his research on rare individuals for whom these capacities are divided in the most bizarre way.

The main part of the brain, known as the *cerebrum*, is made up of two halves—or "hemispheres." The left hemisphere controls the right side of the body and receives information from the right half of the visual field, while the right hemisphere controls the left side of the body and receives information from the left half of the visual field. In most people, information from the two hemispheres is freely shared through the *corpus callosum*, which connects the two hemispheres. However, a radical procedure for treating severe epilepsy involves severing the corpus callosum so that the two hemispheres are unable to communicate. These individuals are colloquially referred to as having a "split-brain."

Sperry and his colleagues conducted a series of experiments involving split-brain patients. Their findings were extraordinary. One of these experiments involved feeding visual information only to one half of the brain. Normally our eyes flit around constantly, meaning that objects in the environment feature in both the left and the right half of the visual field in turn. To avoid this, the patient was instructed to stare at a dot in the center of a computer monitor, so that the left half of the screen would be visible only to the right hemisphere of the patient's brain (which, remember, receives information from the left half of the visual field), and the right half of the screen would be visible only to the left hemisphere (which receives information from the right half of the visual field). If an image appears on the right half of the screen—say, a picture of a piano—then the patient will be able to name the object that has appeared. But if an image appears on the left half of the screen—a picture of a bell, for example—the patient will say that she sees nothing. Why is this? Because, as Sperry confirmed with these experiments, the language center of the person is located in the left hemisphere and the left hemisphere didn't "see" the image of the bell.

Does this mean that the person did not see the bell? It would be natural to assume that, given that the patient reports seeing nothing. However, if the patient is asked simply to draw with her left hand (controlled by the right hemisphere), she will draw the bell! It appears that the right hemisphere, but not the left, experienced the image of the bell. Stranger still, when the drawing of the bell is made visible to the left hemisphere and the patient is asked to explain why she drew a bell, she will formulate some plausible-sounding explanation, e.g., "The piano image made me think of music, and bells are musical." The speech-controlling left hemisphere seems to be confabulating to make sense of a fact about its behavior that is otherwise inexplicable to it.

Another experiment conducted by Sperry's collaborator, Michael Gazzaniga, involved paintings by the sixteenth-century artist Giuseppe Arcimboldo, famous for painting faces made out of a variety of objects, such as fruit, flowers, meat, or books.[19] Gazzaniga set the split-brain subject in front of two buttons, one they could press to indicate they were seeing a face and one to indicate they were seeing fruit (or books, or whichever objects formed the face in the Arcimboldo painting in question). When one of Arcimboldo's paintings was presented in the left half of the visual field (so that the image was seen by the right hemisphere), the patient pressed the "face" button. But when the painting was shown to the right half of the visual field (so that the image was seen by the left hemisphere), the patient pressed only the "fruit" button, indicating that the face had not been visible to her. What this reveals is that the capacity to recognize faces is located in the right hemisphere of the brain.

What on earth is going on here? On the face of it, we seem to have two conscious minds located within the one brain. The "mind" located in the left hemisphere is in charge of speech but is unable to recognize faces; the "mind" located in the right hemisphere has facial recognition but is effectively mute.

The panpsychist philosopher Luke Roelofs, currently a research fellow at Bochum University in Germany, believes that the split-brain cases may provide crucial insights into the secrets of mental combination.[20] The severing of the corpus callosum seems to lead to mental *de-combination:* what was once a single, unified mind, now becomes two separate conscious subjects. If that's right, then we may be able, by imagining the reverse of what led to de-combination, to infer what is required for mental combination. In this way, split-brain cases can potentially give us an empirical handle on mental combination.

Moreover, Roelofs sees in the split-brain cases a way of *re-imagining* the combination problem. Rather than conceiving of the split-brain patient as two minds in one cranium, Roelofs thinks of the patient as *a single individual with disunified consciousness.* Right now, you are enjoying unified conscious experience: your visual experience of these words, your auditory experience of the sounds around you, the tactile experience of the chair beneath you, are all enfolded in a single unified experience. In contrast, although the split-brain patient has both the visual experience available to the left hemisphere (e.g., the image of the piano) and the visual experience available to the right hemisphere (e.g., the image of the bell), she does not enjoy a single unified experience involving each of these experiences (piano + bell). The consciousness of the split-brain patient is fragmented.

For a person with normal, unified consciousness, it is impossible to imagine having disunified consciousness. I will never know what it's like to be a bat because I cannot adopt the perspective of a bat. Similarly, I cannot fully know what it's like to be a split-brain patient, as I cannot adopt the perspective of someone whose consciousness is fragmented into isolated pockets. Nonetheless, the idea of disunified consciousness does not seem to be contradictory, and overall this may be the best way to describe the peculiar reality of split-brain cases.

While the consciousness of the split-brain patient is divided in two, there is still deep unity within each of those two isolated pockets of experience. All aspects of the consciousness of the left hemisphere—colors, shapes, depth perception, etc.—are bound together in unity—as are all aspects of the consciousness of the right hemisphere. But we might imagine the unity of each hemisphere further disintegrating, collapsing into smaller and smaller pockets of separate experience. Assuming the truth of panpsychism, we will eventually, if this process continues, get down to the consciousness of the particles making up the brain. The result: a brain with radically *de-combined* consciousness. Reverse this process, and you've got mental combination.

What kind of brains have radically disunified consciousness? What we are essentially imagining here is the brain of a corpse. For the panpsychist, there is still consciousness in a dead brain in the sense that each of the fundamental particles making it up is conscious. But in the absence of living, cognitive processes going on within it, the consciousness of the particles is not bound together in a single, unified experience. To solve the combination problem, we just need to work out what is going on in the living brain to unify together the otherwise isolated experiences of particles.

Most panpsychists think of the combination problem as:

> How do you get from the consciousness of particles to the consciousness of the brain made up of those particles?

Roelofs has reimagined the combination problem as:

> How do you get from a brain with radically disunified consciousness (i.e., consciousness split up into isolated particle-sized pockets) to a brain with unified consciousness?

It is just this kind of reimagining of a problem that theoretical progress is built on. An intractable problem—*How does particle consciousness become brain consciousness?*—turns into a tractable one—*How does a disunified brain become a unified brain?* The problem is not solved, but possible ways forward open up. It could be the *unified cognitive goals* of the brain, in terms of processing information and representing the environment, that focus its disunified consciousness into unified states of experience. It's very early days, but this is one research path that might ultimately lead to a solution to the combination problem.

CAN QUANTUM ENTANGLEMENT HELP US MAKE SENSE OF MENTAL COMBINATION?

Einstein never liked quantum mechanics. His own theories of relativity were of course pretty weird in their own right, most notably in their implications for the nature of time. But for some reason, he just couldn't handle the weirdness of quantum theory. Many have heard Einstein's famous line, "God does not play dice," to express his discontent with the probabilistic nature of quantum mechanics. But it was not only the randomness that got under his skin. Einstein could not abide the idea of *superposition* (discussed in chapter 2): the thesis that, before we look at them, particles do not have definite locations and velocities but rather exist in superpositions of various locations and velocities. Despite the great experimental success of quantum mechanics, Einstein was convinced that it can't be the complete story. He yearned for a theory to supplement quantum mechanics that would tell us that particles always have definite velocities and locations even when we're not looking at them.

Einstein formulated various attempts to show that quantum theory couldn't be complete. The most famous was the challenge he put together in 1935, together with collaborators Boris Podolsky and Nathan Rosen. Based on their initials, this became known as the "EPR argument." The EPR argument worked off the fact that the properties of distinct particles could become correlated in certain ways. For example, we can set up a situation in which a single particle breaks into two particles of equal mass shooting off in opposite directions at the same speed. In this situation, we know that, so long as the particles are not interrupted on their path, they will continue to travel at the same speed. Or rather, assuming quantum mechanics, we should say that the two particles will have the same speed *when observed,* as before being observed each exists in a superposition of having a variety of speeds.

Now here's the crucial point. We don't need to observe *both* of the particles in order to transform them from being in a superposition to having a definite speed. Given that their speeds are correlated, if we measure one particle—call it "X"—and find it to have a certain speed—call the speed "S"—we can thereby infer that the other particle—let's call it "Y"—will also have speed S. It's as though observing particle X, resulting in X having speed S, instantaneously impacts particle Y, ensuring that Y also has speed S. What Einstein, Podolsky, and Rosen realized is that this mysterious correlation between the two particles would turn up even if at the moment of observation X and Y are separated by such enormous distances that it would take millions of years for light to travel between them. But according to the theory of relativity, nothing travels faster than light and therefore in such a situation it would take millions of years for X to impact on Y (or vice versa).

What is the conclusion of the EPR argument? Suppose X and Y have been traveling away from each other unobserved for thousands of years and are now located on opposite sides of the galaxy. And suppose I measure the speed of X and find it to be S. I can now know that at exactly that same moment Y also has speed S. But given that Y is too far away for my observation of X to impact on it or for information to travel between the two particles, it surely can't be that my observing X caused Y to have speed S. The EPR argument concludes, therefore, that X and Y must have *always* been traveling at speed S, from the first moment they shot apart from each other. Quantum mechanics can't tell us what speed they were traveling before observation. But that's just because, according to the EPR argument, quantum mechanics is incomplete.

The proponents of quantum mechanics were unconvinced. If our best scientific theory cannot tell us the definite speed of the particles before observation, then there is no sense in believing that they have a definite speed. We simply must accept that the particles are related in such a way that when one of them comes to have a definite speed, the other instantaneously comes to have the same speed. This relationship is an instance of what became known as "quantum entanglement," technically defined as a situation in which the quantum state of one particle cannot be described independently of the quantum state of another.

For thirty years this seemed like an irresolvable dispute, with each side taking what was effectively a faith position. Einstein and like-minded physicists just couldn't believe that particles could stand in such mysterious relationships to each other; the proponents of quantum mechanics insisted on taking quantum theory at face value. There seemed to be no way to test which hypothesis was correct. After all, quantum mechanics tells us that particles cease to be in superpositions the moment we observe them. So how on earth could we ever know whether or not the particles had a definite speed before we took the trouble to measure one of them?

In 1964, nearly a decade after Einstein's death, the physicist John Bell realized that there was in fact a way to test which side was correct. Bell's insight is beautiful and a real testimony to the importance of deep thought in science. In the early 1970s we developed the technology to perform Bell's experiment. The result was that quantum theory was vindicated and the EPR argument was shown to be unsound. Since then, quantum entanglement, crazy as it sounds, is one of the best-supported facts of modern science. Particles really can act as one even when light-years apart.*

In the next section I'll explain Bell's experiment. But if you'd rather just take my word for it that there's strong evidence for quantum entanglement, you can skip to the section after where I discuss whether quantum entanglement can help with the combination problem.

* There is an interpretation of quantum mechanics, namely Bohmian mechanics, which is consistent with Bell's findings but denies that there are superpositions. After I have explained what I take to be the connection between entanglement and the combination problem, I will have a further note on how Bohmian mechanics bears on the discussion.

BELL'S EXPERIMENT

The imagined experiment of Einstein, Podolsky, and Rosen discussed above focused on the velocity of particles. The physicist David Bohm demonstrated that a similar experiment might be conducted focusing on the property of particles known as "spin." Physicists have known since the 1920s that a particle spins around its axis. Although, as you might expect, spin at the quantum level is not entirely analogous to the spin of a basketball on a player's finger. For one thing, in contrast to the macroscopic case, the spin of a particle cannot be measured around more than one axis simultaneously. Indeed, one's choice as to which axis to measure seems to determine the axis about which the particle spins. Choose whatever axis you like, and you'll find the particle spinning around that very axis in one of two directions: "up" or "down."

We don't need to worry too much about the intriguing peculiarities of spin. What's crucial for our purposes here is that quantum mechanics tells us that the spins of particles can become entangled. So we can have two particles, X and Y, such that if we measure the spin of X around a certain axis and it turns out to be *up*, then we know that the spin of Y around that same axis must be *down*. (Note that in our previous example, the properties of the entangled particles always ended up being the same, while in this example, the properties of the entangled particles always end up being opposite.)

Building on the work of David Bohm, Bell imagined the following experiment. We send two entangled particles off in opposite directions to two machines at great distances from each other. The machines are set up to measure the spins of the particles around one of three axes, where those three axes are separated from each other by 120 degrees. When the particle reaches the machine, the machine chooses at random one of the three axes to base the measurement on. If it so happens that both machines choose the same axis, and one measures its particle to be *up*, we can infer that the other will measure the spin of its particle to be *down*. The experiment is repeated many times.

Here's the crucial question: How often will the machines measure the particles to have the same spin? We won't go into the mathematical complexities here, but the equations of quantum theory tell us that the machines will record the same spin 50 percent of the time. What Bell realized is that the hypothesis of Einstein, Podolsky, and Rosen gave a different answer. Nobody denies that the spin properties of the two particles are correlated: if X ends up being measured to have *up* spin around a given axis, then Y must have *down* spin around that axis (and vice versa). But Einstein, Podolsky, and Rosen believed that that correlation must have been decided when the particles were together (they doubted it could be decided when they're measured, as at that point the particles are too far apart for a signal to pass between them). With just this assumption in place, we can work out the different possible combinations of spins for the two particles. Here are two possible combinations:

- *First Possible Combination*—X is spin *up* in the 1st and 3rd axes and spin *down* in the 2nd, while Y is spin *down* in the 1st and 3rd axes and spin *up* in the 2nd.
- *Second Possible Combination*—X is spin *up* in all three axes, while Y is spin *down* in all three axes.

There are many other possible combinations; if you have some time on your hands, you can work them all out.

For each possible combination, we can consider different possible random choices of which axes to measure, and then work out with reference to each of these possible choices whether or not the spins of the particles would match. For example, focusing on the first possible combination listed above, here are two possible outcomes:

- Both particles are measured along the 1st axis, in which case they will be measured to have *different spins*.
- X is measured along the 1st axis and Y is measured along the 2nd axis, in which case they will be measured to have the *same spin*.

Again, if you have some time on your hands, you can work out all the possible outcomes. If you do, you'll find out that it's slightly more likely than not that the particles will be measured to have the same spin.

As a result, the two hypotheses end up having different predictions:

- Quantum theory predicts that the two machines will measure the same spin 50 percent of the time
- The hypothesis defended by EPR predicts that the two machines will measure the same spin more than 50 percent of the time.

It is the former prediction which has now been confirmed experimentally.

BACK TO THE COMBINATION PROBLEM

It is commonly assumed that the physical world is made up of little things. This leads many to adopt what I call the "Lego brick" view of reality: You take lots of little things—usually conceived of as fundamental particles—stick them together, and you get big things. If we want another "ism," we could call this view "micro-reductionism." According to micro-reductionism, all of the richness and variety of nature is reducible to the properties and arrangements of fundamental particles. From tables and chairs to planets and stars, all are just complex arrangements of tiny, micro-level particles.

We can perhaps make micro-reductionism more vivid with reference to an intriguing thought experiment formulated by the eighteenth/nineteenth-century French physicist Pierre-Simon Laplace:

> We may regard the present state of the universe as the effect of its past and the cause of its future. An intellect which at a certain moment would know all forces that set nature in motion, and all positions of all items of which nature is composed, if this intellect were also vast enough to submit these data to analysis, it would embrace in a single formula the movements of the greatest bodies of the universe and those of the tiniest atom; for such an intellect nothing would be uncertain and the future just like the past would be present before its eyes.[21]

I'm not sure why, but the super-intelligence being speculated about in the above passage has become known as "Laplace's demon." In any case, as well as being a vivid expression of causal determinism—the view that everything that happens in the universe is causally determined by prior events—Laplace's thought experiment serves to illustrate the idea of micro-reductionism. Merely by knowing everything about the most basic constituents of the physical world, and the relationships between them, Laplace believed that the demon would be able to work out everything else there is to know about reality. The demon would know, for example, who won the World Cup in 1966, what really happened on the *Mary Celeste*, and how many grains of sand there are in the Gobi Desert.

Many have micro-reductionism in mind when they pose the combination problem for panpsychism. It is assumed that my brain is nothing more than an incredibly complex arrangement of particles, and therefore, if panpsychism is true, the consciousness of my brain must be nothing more than the consciousness of the particles that make it up. Panpsychists like Luke Roelofs attempt to formulate a solution to the combination problem consistent with micro-reductionism. They imagine that if Laplace's demon knew everything there was to know about the consciousness of the particles making up my brain, and the relations between them, she would be able to deduce what my consciousness is like. We might call this kind of approach "reductionist panpsychism."

Many take micro-reductionism to be supported by science. Perhaps this is true of the classical physics of the nineteenth century. However, the experimental confirmation of quantum entanglement that finally came in the 1970s gives us overwhelming reason to think we live in a world in which micro-reductionism is false. The equations governing the behavior of a system of entangled particles in superposition govern *the whole system* rather than its individual parts. Even if Laplace's demon knew all the facts about each particle, she still wouldn't know everything there is to know about the system as a whole. A system of entangled particles is *more than the sum of its parts.**

While few deny the reality of quantum entanglement, some argue that micro-reductionism is true at chemical and biological levels. But it is not clear that this commonly held belief has observational support. As it happens, a couple of my philosophy of science colleagues at the University of Durham are two of the most prominent opponents of micro-reductionism. My colleague Robin Hendry has presented a strong empirical case against the reducibility of chemistry to physics.[22] If Hendry is right, Laplace's demon wouldn't even be able to work out the laws of chemistry. And, in her book *How the Laws of Physics Lie*, Nancy Cartwright argues that across the board scientists are too quick to generalize what is discovered about matter from the highly controlled environments of laboratories and experiments. Even if micro-reductionism is true in many of the very simple systems that are studied under laboratory conditions, it does not follow that it is true in highly complex biological systems. Overall, Cartwright argues, the empirical evidence supports a messy "patchwork" world, in which different complex systems have their own distinctive emergent causal capacities.

In any scientific paradigm, there are certain convictions that are held on the basis of evidence, and there are dogmas which are simply accepted as part of the zeitgeist. I suspect the belief in micro-reductionism may be an instance of the latter. It is something that many scientists take for granted and yet there

* In the Bohmian interpretation of quantum mechanics, particles do not form wholes that are more than the sum of their parts. However, Bohmian mechanics postulates both particles and wavelike entities, and the latter are not reducible to facts about micro-level entities. So even on the Bohmian view we live in a world in which micro-reductionism, understood as a theory about the whole of reality, is false.

are no peer-reviewed scientific papers demonstrating that, say, micro-reductionism is true of a complex living system like the brain.

This raises the question: Why on earth are we trying to solve the combination problem under the pretense of micro-reductionism, when we know that micro-reductionism is false at the quantum level and we have no good reason to think it's true at other levels? Indeed, some panpsychists—we can call them "emergentists"—reject the assumption of micro-reductionism. On this approach to panpsychism, conscious systems in brains are, like entangled systems, more than the sum of their parts. The emergentist panpsychist tries to solve the combination problem not by trying to make sense of how lots of little conscious entities somehow "add up" to a big consciousness, but rather by trying to discover the basic principles of nature which give rise to *emergent wholes,* that is to say, complex systems that are more than the sum of their parts.

One of the leading proponents of emergentist panpsychism is Hedda Hassel Mørch, currently a research fellow at the University of Oslo.[23] Mørch distinguishes two forms of emergence: intrinsic and extrinsic. Extrinsic emergence is the emergence of new forms of behavior or causal capacity. Suppose, as Nancy Cartwright believes, the causal capacities of biological systems cannot be wholly explained in terms of the causal capacities of their parts; this would be an example of *extrinsic* emergence. *Intrinsic* emergence, in contrast, is the emergence of new forms of conscious intrinsic nature, which may or may not have a distinctive behavioral manifestation. The behavior of a complex system may be entirely predictable from the basic laws of physics, even if at the macro-level its intrinsic nature involves an emergent form of consciousness not found at the level of fundamental particles. Mental combination, according to emergentist panpsychism, is a form of intrinsic emergence.

Mørch's work tries to marry emergentist panpsychism with the Integrated Information Theory of consciousness (or IIT, discussed in chapter 2 and above). Rather than trying to analyze complex systems into their component parts, the emergentist tries to map out the features of complex systems that mark them out as emergent wholes. In the context of IIT, three features are relevant:

Information

This book contains a fair bit of information. If nothing else, it has informed you of certain prominent views in contemporary philosophy of consciousness. But most of the information in this book concerns things outside of the book; Einstein and Eddington, for example, are not literally contained within the pages of the book. Furthermore, the fact that this book holds information is dependent on human linguistic conventions. The words of this book have no meaning independently of the conventions of the English language.

In contrast, as Mørch explains in her work, IIT is interested in the information a system contains *about itself*, and that it has independently of human conventions.[24] Information in this sense is a matter of how much the system in question *constrains its own past and future possibilities.* The brain has a great deal of information, as there are relatively few possible states (of the brain) in its immediate past and future that are compatible with its present state (at any given moment). The retina of the eye, in contrast, possesses relatively little information, because at any given moment there is a huge range of possible states (of the retina) in its immediate future, depending on what sensory stimulus it receives from the external environment.

Integration

Integration is a measure of how much the information in a complex system is dependent on the interconnections between the parts of the system. Again, Mørch offers a contrast between brains and books. As well as being relational (dependent on human conventions), the information in this book is not very integrated. If, heaven forbid, you tore a page out of this book, you would not thereby destroy any information. The rest of the book plus the missing page would contain as much information as the original book. Even if you tore the book in half, you would end up with two halves each of which contains half of the information the original book contained.

The brain is quite different. Each neuron is connected with around ten thousand other neurons, and the brain's informational structures are highly dependent on these intricate connections. If you severed the connections around one bit of the brain, or cut the brain in half, a huge amount of information would be lost. This marks a crucial difference between computers and brains. A computer could in principle contain as much information as a brain, but the degree of information is not strongly determined by its connectivity. Today's computers are modular systems with feed-forward connectivity, and each transistor is connected to only a few others. For these reasons, severing a part from the rest of the system—like tearing out a page of a book—will not massively reduce its informational content.

Maximality

Almost everything contains *some* integrated information. But IIT does not hold that literally everything has consciousness. It does not predict, for example, that a random collection of pebbles on a beach has its own conscious awareness. According to IIT, a system is conscious when it is a *maximum* of integrated information, a notion we can define as follows:

System S is a maximum of integrated information when the following two principles are satisfied:

- *Not surpassed from below:* No proper part of S has more integrated information than S.
- *Not surpassed from above:* S is not a proper part of something with more integrated information than S.

We can make this clear with some concrete examples:

- *An example of being surpassed from above:* A single neuron has a fair bit of integrated information. However, a single neuron is not a maximum of integrated information, as it is *surpassed from above:* the brain in which it is contained has much more integrated information than the neuron itself.
- *An example of being surpassed from below:* A human society has a great deal of integrated information, due to its complex social connections. However, a society is not a maximum of integration, as it is *surpassed from below:* people make up societies, and their brains have significantly more integrated information than does the society as a whole.

A brain is plausibly a maximum of integrated information, as it neither contains nor is contained within something with a greater level of integrated information. Having said that, it is interesting to note that the connectivity of human societies is vastly increasing due to the internet. On average, any two Facebook users are separated from each other by only 3.57 other users. The relatively new science of network theory provides rich resources for characterizing these increases in connectivity. If IIT is to be believed, then we should perhaps be wary of the growth of social connectivity. This is because IIT predicts that if the growth of internet-based connectivity ever resulted in the amount of integrated information in society surpassing the amount of integrated information in a human brain, then not only would society become conscious but human brains would be "absorbed" into that higher form of consciousness. Brains would cease to be conscious in their own right and would instead become mere cogs in the mega-conscious entity that is the society including its internet-based connectivity.

For what it's worth, this scenario is oddly reminiscent of that predicted by the heretical Catholic priest and paleontologist Pierre Teilhard de Chardin (1881–1955). At a time when some elements of the Church were hesitant about Darwin's theory, Teilhard de Chardin saw in evolution an inspiring vision of an evolving cosmos. Looking back in time, he saw three big leaps in historical evolution: the emergence of life, of consciousness, and finally of self-consciousness in human beings. Looking forward in time, he believed that the next leap would involve increasing levels of connectivity in human societies across the globe giving rise to a new form of life and consciousness that he called the "noosphere."[25]

In any case, the huge amount of integrated information in our brains means that this is not a possibility that need concern us in the near future. What is of immediate interest is that the presence of maximal integrated information may be the marker of emergent consciousness. This is the proposal that Mørch is exploring in the context of Russell-Eddington panpsychism.

What is distinctive about the emergentist panpsychist project is that it doesn't try to analyze human and animal consciousness in terms of more basic forms of consciousness. Instead, the emergentist panpsychist simply postulates basic principles of nature, perhaps those expressed by IIT, in virtue of which higher-level forms of consciousness come into being. Some philosophers confuse emergentist panpsychism with dualism, due to the fact that both theories commit to basic principles of nature being involved in the production of human consciousness. But there is a crucial difference between the two views. The dualist's "psycho-physical laws" (see chapter 2) link physical happenings in the brain with nonphysical happenings in the immaterial mind. The problem with this view, as we explored in chapter 2, is that neuroscience shows no sign of such interactions between the physical and the immaterial. The fundamental principles of nature appealed to by the emergentist panpsychist, in contrast, link micro-level consciousness-involving processes in the brain with macro-level consciousness-involving processes in the brain. In terms of our current knowledge of the brain, there are no empirical grounds for doubting that different levels of the brain are mediated by fundamental principles. Indeed, as Mørch points out, intrinsic emergence may have no obvious behavioral manifestation.*

* In the first footnote of chapter 2 (p. 26), I briefly discussed property dualism, the view that consciousness is a nonphysical property of the brain (rather than of an immaterial soul). Does emergentist panpsychism have any advantage over this position? In my view, property dualism faces pretty much all of the same difficulties that traditional substance dualism (the

WHICH HARD PROBLEM IS THE HARDEST HARD PROBLEM?

Which of these two research programs—reductionist and emergentist—is more promising? The reductionist project still has significant work to do in terms of its theoretical underpinnings. More specifically, the reductionist panpsychist owes us an intelligible account of how complex consciousness might be "built up" from simpler forms of consciousness. Much progress has been made on this issue, but a fully satisfying account of mental combination still evades us. We are used to the distinction in physics between theoretical and experimental practitioners. Until more progress is made on the combination problem, reductionist panpsychism will remain very much in the theoretical branch of consciousness science.

Of course, reductionist panpsychism is not unique in this regard. As we have seen, almost all current theories of consciousness have significant theoretical hurdles to overcome. Each has its own "hard problem":

---

commitment to immaterial minds) faces, as neuroscience shows no sign of the causal effects of nonphysical properties. The emergentist panpsychist avoids these concerns by holding that macro-level forms of consciousness are the intrinsic nature of macro-level brain states, and hence the observable behavior of brain states can be identified with the behavior of macro-level consciousness. Those more familiar with the academic literature will know that these worries about mind-body interaction revolve around the threat of *overdetermination:* that there might end up being multiple sufficient causes of behavior. Mørch avoids this worry by supposing that, in cases of emergence, the emergent whole (rather than the micro-level parts) is the fundamental cause of behavior. I'm inclined to think that we don't know enough about the brain to rule out that the emergent whole and the micro-level parts both partially contribute to determining behavior.

- *The hard problem for materialism.* Materialists have the theoretical obligation to explain how *subjective qualities* could be accounted for in terms of *objective quantities.*
- *The hard problem for dualism.* Dualists have to explain why empirical investigation of the brain shows no trace of mind-brain interaction.
- *The hard problem for reductionist panpsychism.* Reductionist panpsychists must solve the combination problem.*

I have argued that the reductionist panpsychist faces the least hard "hard problem." The reader can decide for herself whether she agrees.

Of all the theories of consciousness, emergentist panpsychism has the fewest problems of a theoretical nature. Indeed, arguably this view faces only "easy" problems, by which I mean problems which we can in principle make progress on empirically.† For the emergentist panpsychist, it is fundamental principles of nature that take us from micro-level consciousness to the consciousness of emergent complex systems. By definition, fundamental principles of nature cannot be *explained*—if they could be explained they wouldn't be fundamental—they can

* Neutral monists, whom we have not discussed in much detail, have the challenge of giving a positive, noncircular characterization of the supposedly "neutral," i.e., neither physical nor mental, intrinsic nature of matter. Sam Coleman, for example, proposes that the intrinsic nature of matter is constituted of *unexperienced qualities.* I discuss these issues in much more detail, and also argue against Coleman's view, in my academic book, *Consciousness and Fundamental Reality.*

† Of course, the "easy" problems are still nightmarishly hard; see chapter 2 for more on the distinction between "hard" and "easy" problems of consciousness.

only be *described*. And hence the emergentist panpsychist can cut straight to the empirical task of trying to formulate and test various candidates for being the fundamental principles that link lower and higher levels of consciousness.

At this stage, emergentist panpsychism is the theory of consciousness most conducive to empirical progress. Having said that, if the pace of theoretical progress on the combination problem continues to be rapid, then we may soon have testable models for both forms of panpsychism.

I am agnostic as to whether the reductionist or the emergentist approach to the combination problem represents the best hope for panpsychism. But I do know that both are vibrant research programs and I have every confidence that progress will be made in the coming decades. A new generation of panpsychists, theorists such as Roelofs and Mørch and many others, are some of the sharpest minds in the contemporary science of consciousness.* Here's a prediction: In twenty years' time, the idea that panpsychism can be quickly dismissed as "crazy" will seem, well, crazy.

A MANIFESTO FOR A POST-GALILEAN SCIENCE OF CONSCIOUSNESS

The quantitative conception of science bequeathed to us by Galileo has been extraordinarily successful. By focusing exclusively on what can be captured in mathematics, scientists have

* Although in general the previous generation was committed to materialism, it does contain some leading figures making invaluable contributions to the panpsychist research program, for example, Galen Strawson, William Seager, Thomas Nagel, David Chalmers (who, although officially a dualist, also takes panpsychism very seriously), and Godehard Brüntrup.

been able to construct mathematical models of nature with ever greater predictive power. These models have enabled us to manipulate the natural world in undreamed of ways, resulting in extraordinary technology. We are now living through a period of history in which people are so blown away by the success of physical science, so moved by the wonders of technology, that they feel strongly inclined to think that the mathematical models of physics capture the whole of reality.

But the purely quantitative science of Galileo cannot capture the qualitative reality of subjective consciousness. If we want a truly complete theory of reality, then we must face up to this inherent limitation of our current scientific paradigm. This does not mean getting rid of physical science, but it does mean incorporating physical science in a more expansive "post-Galilean" science of reality. The aim of a post-Galilean science would be to formulate the most simple and parsimonious theory that is able to account for *both* the quantitative data of physical science—known from observation and experiment—*and* the reality of subjective qualities—known through our immediate awareness of our own experience.

A new generation of theoreticians are already recognizing the need to do this. In the past twenty years, consciousness has gone from being a taboo topic to being accepted as a "hard problem" for science. However, too often the "hard problem" of consciousness is interpreted as simply a tricky puzzle that will one day go away if we just do a bit more neuroscience. The next stage is for people to see consciousness not as something to be squeezed into the world we already know about from science, but as an *epistemological starting point* on a par with the epistemological starting points we get from observation and experiments. Consciousness is not a "mystery"; nothing is more familiar. What is mysterious is reality, and our knowledge of consciousness is one of the best clues we have for working out what that mysterious thing is like.

The basic commitments of this new research program can be summed up as follows:

The Post-Galilean Manifesto

- *Realism About Consciousness:* The reality of subjective consciousness is a basic datum in its own right, equal in status to the data of observation and experiment.
- *Empiricism:* The quantitative data of observation and experiment are foundational, equal in status to the qualitative data of consciousness.
- *Anti-Dualism:* Consciousness is not separate from the physical world; rather consciousness is located in the intrinsic nature of the physical world.
- *Panpsychist Methodology:* We should aim to account for human and animal consciousness in terms of more basic forms of consciousness, basic forms of consciousness which are postulated to exist as basic properties of matter.*

Within this unified research program there are two camps: reductionists and emergentists. Their immediate goals are as follows:

* I consider the first two commitments sacrosanct but am more flexible on the latter two. Neutral monists, such as Sam Coleman, Tom McClelland, Susan Schneider, and Daniel Stoljar, accept the first three but not the fourth. Naturalistic dualists, such as David Chalmers and Martine Nida-Rümelin, accept the first two but not the latter two. I am happy to consider all of them comrades in the post-Galilean revolution.

- *Reductionist panpsychists*—To solve the combination problem, by giving a general account of how complex forms of consciousness can be built up from simpler forms of consciousness.
- *Emergentist panpsychists*—To formulate and test theories of the basic principles of nature underlying the emergence of higher-level forms of consciousness from more basic forms of consciousness.

Progress on these goals will not be achieved by philosophers alone. It will require philosophers and physical scientists working in collaboration. Just as it took a century working within the Darwinian paradigm to get to DNA, so it will require decades or centuries of interdisciplinary labor to fill in the details of Russell-Eddington panpsychism. The problem is that this collaborative project makes no sense in our current scientific paradigm, according to which consciousness, if it exists, must be accounted for in the purely quantitative language of physical science.

The problem of consciousness will not be solved within the Galilean paradigm. We must move to a post-Galilean paradigm, in which the data of consciousness and the data of physics are both taken seriously. Nothing less than a revolution is called for, and it's already on its way.

TECHNICAL APPENDIX B: WHY DO WE NEED INTRINSIC NATURES?

I talked in the preceding chapter about physical science telling us "only what things do," arguing that because of this physical science can never give us the complete truth about the physical universe, even when we set aside the problem of consciousness. A common response to this concern is:

> Why think there's more to matter than what it does? If physics just tells us what matter does, then maybe that's all there is to matter. Maybe once you know what an electron *does,* you know everything there is to know about what an electron *is.*

On this view—known in the philosophical literature as "causal structuralism"—physical entities are not so much "beings" as "doings." Causal structuralism takes a bit of getting used to, but many would argue that it is coherent.

The problem is that when I said that the equations of physics "tell us what matter does," this was really just a loose way of saying that they are a tool for prediction. In fact, careful reflection reveals that physical science doesn't even tell us what matter does.

Let us take a simple example. The mass of two objects creates a force between them, which all things being equal causes them to attract, i.e., to lessen the distance between them. It seems at first like the previous sentence is telling us what mass does. But to really understand the causal impact of mass, we would need to know what "force" is and what "spatial distance" is. Of course, we recognize the presence of these things in our experiments as well as in ordinary experience. But the equations of physics don't tell us what the reality of these phenomena consists in. Rather they characterize them in terms of physical properties like "mass," the phenomenon we began with. In other words, we cannot understand what intrinsic physical properties like "mass" and "charge" are until we know what "force" and "distance" are—because the former are defined in terms of the latter—but until we know what "mass" and "charge" are we cannot know what "force" and "distance" are—because the latter are defined in terms of the former. The buck is continuously passed and an explanation of what anything is, or even what it does, is never given. This is the *circularity objection* against causal structuralism.

We have noted that Einstein gives a deeper explanation of gravitational attraction, but the vicious circle reemerges on his account too. According to general relativity, mass and spacetime stand in a relationship of mutual causal interaction: mass curves spacetime, and the curvature of spacetime in turn affects the behavior of objects with mass. What is mass? For a causal structuralist, we know what mass is when we know what it does, i.e., when we know the way in which it curves spacetime. But to really understand what this reality amounts to, as opposed to merely being able to make accurate predictions, we need to know what spacetime curvature is. What is spacetime curvature? For a causal structuralist, we understand what spacetime curvature is only when we know what it does, which involves understanding how it affects objects with mass. But we understand this only when we know what mass is. We find ourselves in a classic catch-22: we can understand the nature of mass only when we know what spacetime curvature is, but we can understand the nature of spacetime curvature only when we know what mass is. G. K. Chesterton said, "We cannot all live by taking in each other's washing." Bertrand Russell played on this idea in articulating this worry about circularity: "There are many possible ways of turning some things hitherto regarded as 'real' into mere laws concerning the other things. Obviously there must be a limit to this process, or else all the things in the world will merely be each other's washing."[26]

As I was keen to stress in the main part of the chapter, this is not a criticism of physics itself. We can recognize physical properties in our experience—I may not know what "distance" is, but I know that there's less of it between London and Liverpool than there is between London and Cairo—and by applying the equations of physics to them we can predict the future. This is what physics aims to do, and it does it very well. But if what we want is a *theory of reality*, rather than a tool for prediction, we must be able to give a noncircular characterization of the properties postulated by our theory.

Causal structuralists often contend that the circularity objection begs the question (in the technical sense of assuming in a premise of your argument the very thing you are trying to prove) by assuming that the definition of a given property, such as mass, must be given in isolation from all other properties. Causal structuralism, they say, implies a kind of holism, whereby the nature of a given thing cannot be understood in isolation from all other things. And so the very demand for an account of "mass" in isolation from "spacetime curvature" would seem to be premised on the assumption that causal structuralism is false. If causal structuralism is true, then mass and spacetime curvature (and everything else) must be defined "all at once."*

I agree that mass can be uniquely identified—as it were, homed in on—in terms of its place in the abstract pattern of causal relationships realized by the entire network of physical properties. But doing this doesn't tell us what mass *does*. And according to causal structuralism, physical properties are defined in terms of what they do: by the effect they have on the concrete physical world. Thus, if causal structuralism is true,

* This kind of holistic definition can be done with the help of a logical tool known as a "Ramsey sentence."

in order to know the nature of a physical property, we need to know what that property does and not merely its place in an abstract pattern of causal relationships.

This is all getting very abstract, so let's take a ludicrously simple example. Suppose I have three matchboxes, and I tell you the first contains a "SPLURGE," the second a "BLURGE," and the third a "KURGE." You innocently ask me, "Oh really, what's a SPLURGE?" I answer, "A SPLURGE is something that makes BLURGES." Now, you can't really understand my answer until you know what a BLURGE is, so naturally your next question is, "Fine, so what's a BLURGE??" I respond, "Oh, that's easy, a BLURGE is a thing that makes KURGES." But, in a similar way, you can't understand this answer until you know what a KURGE is, and so—starting to get a bit irritated—you now demand to know: "What on earth is a KURGE???!!" My response: "It's something that makes SPLURGES."

You could be forgiven at this point for deciding you didn't want anything more to do with this conversation. For although the discussion has taught you something about the abstract pattern of causal relationships that exists between SPLURGES, BLURGES, and KURGES, it has left you none the wiser about what any of them actually does. The same is true, although in a more complex way, of the description of physical reality offered by the causal structuralist. If causal structuralism is true, it is logically impossible to gain understanding of what anything does and hence logically impossible to gain understanding of what anything is. If this isn't an unintelligible view, then I don't know what is.

The panpsychist avoids the circularity objection by giving a noncircular account of the intrinsic nature of matter. The character of a given subjective experience is not defined in terms of anything outside of that experience. The pain I'm currently feeling has an intrinsic character that I'm immediately aware of in having the experience. One can't convey that character to someone who hasn't had the experience. Nonetheless, in having the experience and thereby grasping its character, you have a complete understanding of what it involves. And hence, in principle, the panpsychist can give an account of the nature of a physical property, such as mass, without deferring to some other physical property and thereby getting into a vicious circle or regress. If mass is a form of consciousness, then in specifying the character of that form of consciousness you've thereby specified what mass is.

Here's another way of putting the circularity objection. If every word was defined in terms of other words, then all definitions would ultimately be circular and language could never reach beyond itself. In order to get meaning going, we need to have some primitive concepts that are not defined in terms of other concepts. The concepts of physical science are not primitive but *inter-defined:* mass is characterized in terms of distance and force, distance and force are characterized in terms of other phenomena, and so on until we get back to mass. Our concepts of consciousness, in contrast, are primitive in the required sense: a feeling is not defined in terms of anything other than itself.

This is not to say that there are not problems here, nor that we are anywhere near having a plausible candidate for what form of consciousness mass is (or any of the other basic physical properties). Nonetheless, the point remains that panpsychism does not suffer from the circularity that plagues causal structuralism.

One final point on this issue: Even if the circularity objection fails to undermine the coherence of causal structuralism, the Russell-Eddington view that matter has an intrinsic nature still remains a theoretical option. And if it is possible that matter has an intrinsic nature, then it is possible that this is the home of consciousness. Nor is this just a theoretical possibility with nothing to recommend it. If we are persuaded by the arguments of chapters 2 and 3 against dualist and materialist accounts of consciousness, then the Russell-Eddington approach to consciousness will be sufficiently motivated even if causal structuralism is also a coherent view. The circularity objection to causal structuralism merely provides an additional source of support.

5

Consciousness and the Meaning of Life

I've always been curious about the ultimate nature of Reality. Curiosity comes in many forms. Some are interested in the human sphere: our history, our psychology, the variety of ways in which human culture has manifested itself at different times and places. Others are fascinated by the natural world of plants and animals. Still others explore the abstract realm of numbers and sets. Some people are obsessed with solving problems, from curing cancer to working out how to ease congestion in a city center. Others love to care or to create things of beauty. What keeps me awake at night is wondering what *kind of Reality* we live in.

To give it a name, my passion is for *ontology:* the study of Reality in its most general form. It is often assumed that it is the job of physicists to give us these answers, which is probably why the word "ontology" is not really known outside of philosophy. Why would we need another word for physics? But as we have discovered in earlier chapters, physics is simply not in the business of doing ontology.

The first four chapters of this book have been an essay in ontology. Of course we will never know for sure what the Reality we live in is really like. But I have tried to argue that panpsychism is the most probable hypothesis. In my view, the entry for our world in *The Hitchhiker's Guide to the Many Realities* is likely to read: "A physical universe whose intrinsic nature is constituted of consciousness. Worth a visit."[1]

What I'd like to do now is to move beyond strict ontology and explore the implications of panpsychism for human existence. When we're doing ontology, we should be thinking not about which view we'd most like to be true, but which view is most likely to be true. That's what I tried to do in chapters 1–4. However, as it happens, not only do I think that panpsychism is likely to be true: I also think that it is a theory of Reality somewhat more consonant with human happiness than rival views.

THE CLIMATE CRISIS

Since 1980 the temperature of the planet has risen by 0.8 degrees Celsius, resulting in unprecedented melting of the Greenland ice sheet and the acidification of oceans. In 2015, 175 million more people were exposed to heatwaves compared with the average for 1986–2008, and the number of weather-related disasters from 2007 to 2016 was up by 46 percent compared with the average from 1990 to 1999.[2] This is nothing in comparison to the horrors that await us as temperatures continue to rise. According to recent projections, global temperatures are set to increase by 3.2 degrees by the end of century.[3] This will lock in sea level rises that will mean that the cities, towns, and villages currently occupied by 175 million people—including Hong Kong and Miami—will eventually be underwater.

There is overwhelming scientific evidence that this warming is largely caused by the actions of human beings. Surveys of the scientific literature have consistently found that over 90 percent of scientists believe that climate change is real and man-made, with most surveys asserting a consensus of 97 percent.[4] And yet there is in the public mind a perception that the reality of man-made climate change is uncertain. This is in large part caused by a sustained lobbying effort from the fossil fuel industry aimed at spreading seeds of doubt. But it may also result from a failure to appreciate how uncertain most of human knowledge is. Many believe that science provides "proven facts," and against this assumption any degree of uncertainty can seem to render a hypothesis "unscientific," a matter of speculation rather than demonstrable knowledge.

Philosophy can help with this. David Hume was one of the great skeptics of philosophy. While Hume did not dispute the reality of our conscious experience, he argued that there was no way of demonstrating that our conscious experience corresponds to anything real, e.g., that a conscious experience of a table corresponds to a real physical table out there in the external world. In spite of this, Hume did not think that skepticism was something to be feared:

> . . . though a [skeptical philosopher] may throw himself or others into a momentary amazement and confusion by his profound reasonings; the first and most trivial event in life will put to flight all his doubts and scruples, and leave him the same, in every point of action and speculation . . . with those who never concerned themselves in any philosophical researches. When he awakes from his dream, he will be the first to join in the laugh against himself, and to confess, that all his objections are mere amusement, and can have no other tendency than to show the whimsical condition of mankind . . .[5]

In other words, you merely have to stop philosophizing and get on with life and these skeptical worries about the external world evaporate like morning mist. As Hume put it, "nature is always too strong for principle."

What then is the point of engaging with the philosophical skeptical arguments if we're just going to get on with life and forget about them? The benefit of such reflections, according to Hume, is that they can lead to a healthier relationship with evidence:

> The greater part of mankind are naturally apt to be affirmative and dogmatical in their opinions; and while they . . . have no idea of any counterpoising argument, they throw themselves precipitately into the principles, to which they are inclined; not have they any indulgence for those who entertain opposite sentiments. To hesitate or balance perplexes their understanding, checks their passion, and suspends their action. They are, therefore, impatient to escape from a state, which to them is so uneasy; and they think, that they can never remove themselves far enough from it, by the violence of their affirmations and obstinacy of their belief. But could such dogmatical reasoners become sensible of the strange infirmities of human understanding, even in its most perfect state . . . such a reflection would naturally inspire them with more modesty and reserve, and diminish their fond opinion of themselves, and their prejudice against antagonists.[6]

I am struck by how this eloquent description of dogmatic tendencies rings true today. We are living in an increasingly polarized age in which people run away from uncertainty by bolstering their convictions to the point where no alternative is given the slightest credibility. This kind of obstinacy is simply incompatible with the realization that even our most basic beliefs, such as the belief that that there is an external world or that the universe has existed for more than five minutes, are not known with 100 percent certainty. One of the many values of a philosophical education is that it teaches the importance of doubt.[7]

How does this help with climate change skepticism? Paradoxically, the cure for excessive doubt is doubt of a more radical kind. Conspiracy theories thrive in an environment in which certainty is expected, because this expectation sets up a demand that can never be met. When one realizes that little if anything is known with certainty, even whether one's feet exist, one becomes more comfortable with probabilities that fall short of 100 percent. If you start from the idea that there is a core of scientific knowledge that is known with 100 percent certainty, then something accepted by "only" 97 percent of scientists can seem too uncertain to warrant real commitment. But the skeptical philosopher knows that if she were to wait for certainty she would never form a meaningful relationship for fear of befriending a philosophical zombie. To properly understand the human situation is to appreciate that less than certainty can be enough to trust, to engage. Indeed, a threshold of much less than certainty is very often enough to *demand* belief and practical engagement.

We tend to think of skeptical philosophers as cynically withholding belief until they are shown the truth with certainty. In fact, the philosopher who is truly comfortable with skeptical doubt knows that certainty is too much to ask. The consensus of 97 percent of scientists is more than enough.

Even setting aside the dubious doubts surrounding man-made climate change, our inability to take action against climate change is bizarre. In my home country of Britain, 64 percent of people believe that climate change is real and largely caused by humans, and yet there is little political pressure for action.[8] The numerous international agreements have all been inadequate. The Paris Agreement of 2015 went further than previous agreements, with 196 countries signing up to specific pledges aimed at keeping global temperature rises well below 2 degrees Celsius and ideally to prevent rises above 1.5 degrees. The problem is that, according to Climate Action Tracker, the vast majority of countries in the world are way off meeting even the 2 degrees target.[9] This was the case even before Donald Trump exacerbated this mass failure by pulling the United States out of the Paris Agreement in 2017.

Imagine that we discovered tomorrow that a meteor was on course to hit our planet in fifteen years' time and was set to cause the kind of devastation we know to be associated with climate change. No doubt governments would get together to see if there was any way in which this tragedy could be averted. And if there was an option for doing so—perhaps by blasting the meteorite to smithereens—then you can bet there would be significant resources and political will put into fulfilling this project. And yet when we know that our planet is about to suffer devastating increases in temperature, and we know that there is something we could do about it, human beings seem incapable of rising to the moment.

Could our philosophical worldview be partly responsible for our inability to avert climate catastrophe? The writer and campaigner Naomi Klein places blame at the foot of *mind-body dualism,* or as she puts it, the "corrosive separation between mind and body—and between body and earth—from which both the Scientific Revolution and the Industrial Revolution sprang."[10] The dualist conceives of the natural world as a mechanism lacking in the consciousness that sanctifies human existence. It is therefore something to be exploited rather than revered. In particular, Klein blames the scientist and philosopher Francis Bacon for "convincing Britain's elites to abandon, once and for all, pagan notions of the earth as a life-giving mother to whom we owe respect and reverence (and more than a little fear) and accept the role of her dungeon master."[11]

How can dualism be the problem given that our current scientific paradigm is *materialist* rather than dualist? Although materialism is by and large the official party line of the scientific community, it is not clear that it is the general view of the populace. Indeed, materialist David Papineau has powerfully argued that even among those who are persuaded by the arguments for materialism, it is almost psychologically impossible to actually believe that conscious experiences are physical processes in the brain.[12] Closet dualism is revealed in the tendency for the problem of consciousness to be posed by asking how physical processes "give rise to" or "produce" conscious experiences? If we really believed that conscious experiences just were neurophysiological processes, then there would be no question as to how the latter "give rise" to the former. As we discussed in chapter 3, a single thing cannot give rise to itself, and hence if we are asking why brain processes "give rise" to conscious experiences, this can only be because deep down we are really taking them to be distinct things.

Papineau does not think this undermines the arguments in favor of materialism, which he believes to be overwhelming. He simply takes it to be a peculiar psychological fact about human beings that they can't help thinking dualistically. When our official worldview is that biological systems are mechanistic, most of us end up believing that consciousness is really something over and above those mechanistic biological systems. In other words, we end up being closet dualists.

Dualism can create an unhealthy relationship with nature in at least two respects. Firstly, it creates a sense of separation. Dualism implies that, as an immaterial mind, I am a radically different kind of thing from the mechanistic world I inhabit. Ontologically speaking, I have nothing in common with a tree. There is no real kinship with nature if dualism is true. Secondly, dualism can imply that nature has no value *in and of itself.* If nature is wholly mechanistic, then it has value only in terms of what it can do for us, either by maintaining our survival or by creating pleasurable experiences for us when we take it in with our senses. There is a worry that dualist thought can encourage the idea that nature is to be used rather than respected as something of value in its own right.

It is no surprise that in this worldview the act of tree hugging is mocked as sentimental silliness. Why would anyone hug a mechanism? Superficially, nature may appear beautiful and teeming with living energy, and probably in many encounters with nature we can't help but believe this. But our intellectual worldview tells us that nature is nothing more than a complex mechanism. It is hard to feel any genuine warmth for the natural world so conceived.

Descartes went so far as to believe that animals are mechanisms, although very few dualists these days would agree with him. Almost everybody will accept that many nonhuman animals are conscious. Given our inability to really accept materialism, we are inclined to think that the brain processes of animals "give rise" to consciousness too. In virtue of being conscious creatures, animals also have inherent value (or at least their conscious minds do). But in the dualistic worldview, we conscious creatures—humans and other animals—are profoundly isolated from each other, housed as we are in this unfeeling mechanism of the physical world. The sense of a unified, interdependent ecosystem that comes so naturally to us when we engage with nature does not fit with the dualism that, so long as we construe nature as purely physical, we cannot but believe.

Panpsychism has the potential to transform our relationship with the natural world. If panpsychism is true, the rain forest is teeming with consciousness. As conscious entities, trees have value in their own right: chopping one down becomes an action of immediate moral significance. Moreover, on the panpsychist worldview, humans have a deep affinity with the natural world: we are conscious creatures embedded in a world of consciousness.

We treat other humans not as objects but as sentient centers of value and purpose. We feel their presence when in close proximity, and we instinctively interpret their actions as flowing from their individual agency. Imagine if children were raised to experience trees and plants in the same way, to see the movement of a plant toward the light as expressing its own desire and conscious drive for life, to accept the tree as an individual locus of sentience. For a child raised in a panpsychist worldview, hugging a conscious tree could be as natural and normal as stroking a cat. It's hard to tell in advance the effects of such a cultural change, but it's reasonable to suppose that children raised in a panpsychist culture would have a much closer relationship with nature and invest a great deal more value in its continued existence.

In fact, independently of the arguments for panpsychism contained in this book, there is now a growing body of evidence that plants have sophisticated mental lives. Monica Gagliano, a research associate professor at the University of Western

Australia in Perth, has shown that pea plants can be subject to conditioned learning.[13] Gagliano proved this by replicating the famous experimental work of the nineteenth-century psychologist Ivan Pavlov. Pavlov taught dogs to associate the ringing of a bell with food, by always ringing a bell before feeding them. Eventually, the mere sound of the bell was enough to cause the dogs to salivate.

In order to set up a similar scenario with her pea seedlings, Gagliano put a pea plant at one end of a Y-shaped tube, so that it could grow in either of two directions, left or right. In one direction was the seedling's "food," in the form of blue light. In normal circumstances, the pea seedlings will instinctively grow toward where the light was last present. However, Gagliano tested whether the seedlings could associate the sound of a computer fan with the presence of the blue light, by repeatedly placing the noise at the end of the tube where the blue light was located. Upon repeated trials, she found that just as Pavlov's dogs had salivated at the sound of the bell, so the pea seedlings grew toward the noise of the computer fan. In both cases, a sound that was initially meaningless to the organisms had come to represent dinnertime. This research has led Gagliano to think of the seedlings as subjects of experience: "If the plant is imagining its dinner arriving, based on a simple fan that is associated to the light, then *who* is doing the imagining? *Who* is thinking here?"[14]

Does a tree falling in a forest make a noise if there's no one there to hear it? Why assume the tree makes noise only when it falls? The research of Ariel Novoplansky, from the Ben-Gurion University of the Negev, has demonstrated that plants can communicate with each other in sophisticated ways.[15] Novoplansky's experiment involved putting plants in a series of adjacent pots, with each plant having one root in its neighbor's pot. He then subjected one of the plants to drought. What he discovered was that this information was passed down the series of plant pots through the roots, as revealed by the fact that all of the plants closed their pores to reduce water loss. Closing of pores is generally the action of thirsty plants, but in this case it was the action of perfectly well-watered plants responding to the danger signals of a neighbor several pots along. The plants were even able to retain the information, which prevented them from dying in the drought that Novoplansky subjected the plants to in a later stage of the experiment.

Suzanne Simard, of the University of British Columbia, has taken research into plant communication outside of the lab and into the forest.[16] In the early days of her research, Simard was ridiculed for wanting to investigate tree communication, and she struggled to find funding. It is these same prejudices—cultural associations with "New Age" ideas—that continue to hamper the panpsychism research program, although thankfully less and less as time goes on. Unperturbed, Simard pushed forward with her scientific work, and the results have been extraordinary.

By injecting trees with isotope traces, Simard has shown that there is beneath our feet a complex web of communication between trees, which she had dubbed the "Wood-Wide Web." Communication happens via *mycorrhiza* structures, which connect trees to other trees via fungi. The trees and the fungi enjoy a quid pro quo relationship: the trees deliver carbon to the fungi and the fungi reciprocate by delivering nutrients to the trees. A dense web of connections is formed in this way, with the busiest trees at the center connected to hundreds of other trees.

The mycorrhiza structures allow for a complex system of egalitarian redistribution, with trees with excess carbon passing some on to their neighbors. It is sometimes claimed that in human societies social harmony is possible only when they are united by strong ties of kinship. No such prejudice exists among trees. Even *across species* there exist networks of reciprocal support. In summer, the birch trees help out the fir trees by passing along carbon, especially to the fir trees that are shaded from the sun. In winter, there is reciprocation: when the birches are leafless, the firs provide much needed carbon support.

Having said that, like humans, trees do exhibit preferential treatment for their own young. Simard has shown that the "mother" trees at the center of the network not only give greater amounts of carbon to their own kin, but also send them defense signals which can increase by a factor of four the young trees' survival chances. This intergenerational transfer is particularly pronounced at the point when the mother trees die, as they pass on their wisdom to the next generation.

On the basis of all this, we now know that plants communicate, learn, and remember. I can see no reason other than anthropic prejudice not to ascribe to them a conscious life of their own.

Admittedly, this does have difficult implications for the ethics of vegetarianism and veganism. Many vegans and vegetarians feel that it is wrong to kill or to exploit sentient creatures. But if plants also have sentience, what is there left to eat? These are very hard ethical questions; it may turn out that some killing of sentient life is inevitable if we want to survive ourselves. But accepting the consciousness of plant life means at the very least accepting that plants have genuine interests, interests that deserve our respect and consideration.

Few people are aware of these transformations in our understanding of plant mental life, and many would still probably dismiss the ideas that trees talk as hippie nonsense. But imagine how our children's relationship with nature could be transformed if they were taught to walk through a forest in the knowledge that they are standing amidst a vibrant community: a buzzing, busy network of mutual support and care.

The cultural revolutionaries of the 1960s aspired to a new relationship with nature, one of love, respect, and harmonious coexistence. These aspirations fell flat without an intellectual worldview in which they made sense. Such a worldview—panpsychism—is now intellectually credible. There is every reason to hope that the new science of consciousness will lead to a new covenant with nature. The only problem is we have such little time.*

ARE WE REALLY FREE?

In the course of our lives we naturally conceive of ourselves as *free agents,* as creatures able to consider a variety of rational considerations and to freely choose among them. Suppose I have to decide by tomorrow whether or not to take a new job. I consider the pros and cons. If I take the job, I'll have to work more, meaning more time away from my family. On the other hand, I'll have more money and hence be able to afford a bigger house. Eventually I decide that I would have a better quality of life in a smaller house with more time with my family, and on that basis I turn down the job.

* Just as I was putting the finishing touches to this manuscript, a new report from the United Nations Intergovernmental Panel on Climate Change was released warning that we have only twelve years left to ensure that temperature rises stay below 1.5 degrees. We also learned more about the importance of sticking to this more ambitious target rather than the target of 2 degrees. To take just one example, ice-free summers in the Arctic will come every 100 years with 1.5 degrees of warming, but every 10 years at 2 degrees.

When any of us is in this kind of situation, we feel that the choice is *up to us.* We could go one way or we could go the other, depending on our judgment of the weight of reasons. And when we do make the decision, we believe that it was made for a particular reason. Sarah married Clare because she knew they'd be happy together. Paul became vegan because, in his judgment, meat and dairy consumption is no longer environmentally sustainable. Angela voted Conservative because she didn't think it was fair to make her pay more tax. Any historical or sociological explanation is dependent on the assumption that human affairs are determined by the responsiveness of people to rational considerations. This is not to say that people are perfectly rational. Of course, humans have all sorts of flawed and confused beliefs and twisted motivations, and even when they do have a clear-sighted understanding of what should be done, they very often, through weakness of will, fail to do it. Nonetheless, social events can be understood only by ascertaining the considerations which the actors involved believed, rightly or wrongly, to be reasons for action.

If materialism is true, all such explanations of human affairs are false. People are physical objects, and hence what they do is determined not by considerations of reason, or even by considerations that are taken to be rational, but by mechanical causes. Daniel Dennett's position is that we treat people *as though* they were rational creatures, a pretense he calls "the intentional stance." We rely on the intentional stance only because of the immense complexity of the underlying mechanical causes of behavior; on the whole, it provides a useful rough-and-ready way of predicting what humans will do. But in principle if I had a complete understanding of the "real causes" of your behavior, in terms of fundamental physics, I could work out much more accurately what you are going to do and dispense with the idea that you are responding to rational considerations (we are reminded of Laplace's demon from chapter 4). Thanks to the beneficence of natural selection, human beings generally behave in ways that are rationally appropriate given the goal of survival, but their apprehension of reasons plays no part in the explanation of what they do. The ultimate causes of behavior are utterly irrational forces operating at the level of electrons and quarks.*

I fully accept that the sense that we are free agents may turn out to be an illusion. The reality of consciousness is a basic datum in a way that the reality of free will is not. I am more certain of the reality of my feelings and experiences than I am of anything else. But anything outside of my immediate conscious awareness is subject to doubt. I have no way of knowing for sure that my sense that I am the ultimate cause of (some of) my actions accurately reflects how things really are. Nonetheless I am unpersuaded by the philosophical and scientific arguments that have been put forth to try to show that free will *is* an illusion.

One of the most popular philosophical arguments against the coherence of free will starts from the claim that there is no middle way between determined behavior on the one hand and completely random and senseless behavior on the other. Either

* Many philosophers, including Dennett in *Freedom Evolves*, pursue the middle-way option of *compatibilism*, according to which human freedom is entirely consistent with a determined universe. I am persuaded by the arguments of E. J. Lowe (in my opinion, one of the finest philosophers of the twentieth and twenty-first centuries) in his book *Personal Agency* that genuine responsiveness to rational considerations is incompatible with the thesis that our actions are compelled by prior events. In any case, one is led to compatibilism only when one is persuaded there is strong scientific reason to give up on the idea that we are free in the way we ordinarily take ourselves to be. As I explain below, I don't believe that science does, at this moment, provide us with grounds for jettisoning our ordinary notion of freedom.

my actions have prior causes, in which case they are determined and so not free in the relevant sense, or my actions "just happen" in a completely arbitrary and uncontrolled way. But arbitrary and uncontrolled actions are not free in any interesting sense either. Genuinely free choices would need to be neither determined nor random, and nothing—it is argued—can possibly avoid both of these extremes.[17]

The problem with this argument is that there is a middle way between determined actions and random choices, or at least there could be. Free choices can be distinguished from random events by the fact that they involve *responsiveness to rational considerations.*[18] Suppose my decision not to take the job was not determined by prior causes. What makes it the case that my decision was not a random event that "just happened" for no reason? The answer is that my decision was made for a reason: in making the decision I was responding to the fact that not taking the job would give me more time with my family. Contrast with the case of a truly random event, such as a sample of the radioactive substance thorium-234 decaying within 24.1 days. Like my decision, this event was not determined in advance: there was a 50 percent chance that the substance would decay and a 50 percent chance is wouldn't. But unlike my decision, the decay of the thorium-234 did not involve responsiveness to rational considerations. It is this that makes it random—an event that "just happened"—as opposed to a free decision.

At this point, free will skeptics will typically demand to know what explains the fact that I decided one way rather than the other. The consistent believer in free will must hold that nothing explains it. Everyone takes some facts as basic and unexplained. Some people take the laws of physics as an unexplained starting point; others the existence of God; others the laws of logic and mathematics. I take the reality of consciousness as a fundamental starting point. The demand to explain *absolutely everything* would create either an infinite regress or a vicious circle.* So we know that some things have to be left unexplained. Why not certain decisions of free agents? Of course, the *action that results* from the decision is not unexplained: it was caused by the free decision. But the decision itself might be a basic fact about reality, not explained in terms of prior causes or anything else. Despite being unexplained, such decisions are not random because they involve responsiveness to reasons. When an agent acts freely, she acts for a reason.

In other words, this common argument against free will blurs together two very different objections:

- *Objection A*—Free choices cannot be coherently distinguished from random events.
- *Objection B*—Free choices are unexplained events.

The appearance of a powerful argument is given by subtly hopping between these two different concerns. But once they are clearly distinguished, it is clear that each can be answered:

- *The Solution to Objection A*—Free choices differ from random events in that they involve responsiveness to rational considerations.
- *The Solution to Objection B*—Everyone has to accept some things as unexplained. Accepting free choices

* I try to show in Technical Appendix B that, in fact, one cannot take physics (on its own) to be the complete and final story of reality, as this leads either to a vicious circle, given that physical terms are interdefined. However, I set these arguments aside here, as, at this stage, I am simply trying to make the more general point that one has to take *something* as basic.

> as unexplained is no less coherent than accepting the big bang as unexplained.

I can see no philosophical objection to the coherence of free will. But just because free will is coherent, this doesn't mean it really exists. After all, dragons and unicorns are logically coherent, but you're not in danger of bumping into them anytime soon. The free will skeptic could accept that free will is coherent while arguing that we have scientific grounds for thinking that it doesn't exist.

The scientific case against free will is most associated with experiments conducted by Benjamin Libet in the 1970s.[19] These experiments involved participants being asked to make a decision to execute some trivial task, such as flexing a hand or pressing a button, at a random moment within a fixed period of time. What Libet wanted to do was to assess:

A. At what time the conscious decision was made to perform the action.
B. When the brain first initiated the activity that led to the action, an event referred to as the "readiness potential."

He assessed (A) by sitting his participants in front of an oscilloscope timer and asking them to notice when they were first aware of the wish or urge to act. To assess (B) he attached EEG electrodes to the participant's scalps in order to record electrical activity involved in the initiation of the action. The shocking result of the experiments was that, on average, the reported time of the conscious urge occurred up to 300 milliseconds after the brain first initiated the action. Many have taken this to be clear evidence that our sense that our conscious minds initiate action is an illusion: in reality it is nonconscious happenings in our brains that determine what we're going to do.

The Libet experiments are fascinating and important. However, the common claim that they prove the nonexistence of free will is far too hasty. Some, including Daniel Dennett, have argued that Libet ignored a far more straightforward interpretation of the data: the participants systematically misjudge the time of the conscious urge by on average 300 milliseconds.[20] After all, the setup of the Libet experiments is highly artificial: subjects are asked to make a decision while simultaneously noting the time at which the decision was made. Perhaps human beings are simply incapable of performing these two tasks at once. However, later experiments using functional magnetic resonance imaging (fMRI) have concluded that the outcome of a decision can be encoded in brain activity of the prefrontal and parietal cortex up to ten seconds before it enters awareness.[21] Such a significant gap strains the credibility of this alternative interpretation of the data.

The deeper problem with Libet's experiments, as well as similar experiments carried out more recently, is that the kind of "decisions"—if indeed they deserve that name—they focus on are not the kind of decisions the defenders of free will are concerned with. Proponents of free will are keen to preserve our natural conviction that we are able to freely respond to rational considerations. But the "choices" Libet set to his participants were completely random and senseless, and the same is true of more recent experiments. There is nothing that speaks in favor of flexing one's arm at a slightly earlier time rather than a slightly later time, or of pressing the button on the left rather than the button on the right. At best Libet's experiments prove that the conscious mind is unable to initiate a completely random and meaningless action. But such an action would not be a case of free choice anyway—not in any sense that we care about—and so this result does nothing to undermine the case for genuine freedom.

To really refute the existence of free will, experiments would have to focus on genuine choices, such as whether to take a job or get married. And even then, we need not suppose that there is a single specific moment at which the decision took place. A long period of deliberation may involve pushing oneself bit by bit toward a certain outcome, an outcome which eventually becomes inevitable. It is hard to see, practically speaking, how one might conduct an experiment to determine whether such decisions are wholly determined by nonconscious prior causes. But this does not give us permission to exaggerate the conclusions of the experiments we are able to do.

Some thinkers hope that quantum mechanics may make room for free will. However, while it is true that there is indeterminism in standard interpretations of quantum mechanics—later events are not rendered inevitable by earlier events—the indeterminism in question is highly circumscribed. Earlier events, according to quantum mechanics, determine the *objective probability* of later events. Genuine free choice is inconsistent with such objective probabilities. Suppose that prior events in my brain make it 75 percent likely that I'll take the job. It follows that if there were a million exact physical duplicates of me in the same position, 750,000 of them would decide to take the job and 250,000 of them would not. Each of these duplicates would feel as though they are making a decision which is utterly under their control. But this can't be so. For if each agent were able to freely choose, then we wouldn't be able to predict how many would choose one way and how many the other, whereas quantum mechanics tells us we can do this. To be genuinely free, the past must not compel me, either by making my choice inevitable or by determining a precise probability of which choice I will make.

I believe it is still an open question whether or not free will is an illusion. But it is worth noting that panpsychism, unlike materialism, is in principle able to preserve its reality. The materialist assumes that everything that happens in the physical world has a prior cause, which either determines precisely what will happen or—in the case of quantum indeterminacy—determines the objective probability of what will happen. But for the panpsychist, a different model is possible. It could be that past events *pressure* physical entities in the present to behave in a certain way but that it is always up to present physical entities whether or not to accept that pressure.

That last statement might seem a little obscure. What I am trying to capture is how things seem to be when I am making a free decision. It feels like it's genuinely up to me, while at the same time there exist certain pressures in the form of my inclinations. To the extent that I crave cheese—the biggest challenge to my ongoing struggle to be vegan—it will be harder for me to choose a vegan lifestyle. In extreme cases, such as torture, the pressure may overpower me, making it inevitable that I will choose in the direction I am being pressured. But in many ordinary cases, it is up to me whether or not to yield to the pressure of my inclinations, or which inclination I choose to yield to.

Common sense tells us that something *similar,* although not quite *the same,* happens in babies and nonhuman animals. Such creatures are not capable of rational deliberation. But it doesn't follow that their inclinations *compel* them to act in certain ways. It is more natural to suppose that their inclinations *pressure* them to act in certain ways. It might even be inevitable that animals will follow those inclinations, in the absence of rational deliberation to counter them. This kind of inevitability is not to be confused with determinism of a kind that is incompatible with freedom. What is important for freedom is that one is not compelled to act by prior causes, as acting through compulsion is inconsistent with genuine responsiveness to rational considerations. However, if the balance of rational considerations clearly favor action A, and a given agent has no desire to act to the contrary, then it may be inevitable that the agent will choose A. The fact that a certain decision is inevitable—perhaps because it's a "no-brainer" what ought to be done—does not entail that it was compelled.

As we move to simpler and simpler forms of life, behavior may be more and more predictable. Indeed, it could be that even complex animals are in principle entirely predictable, at least for Laplace's super-intelligent demon (see chapter 4) if not for ourselves. Again, this doesn't mean the animals in question are compelled to act. It could be that the animal chose the behavior in response to its inclinations, even if—in the absence of rational deliberation—it was inevitable that the animal would follow its strongest inclinations.

It seems to me coherent to suppose that this model also holds in the subatomic realm. Clearly particles don't rationally deliberate, and hence don't "choose" in the sense that human beings "choose." But it could be that they act through responsiveness to inclinations, inclinations produced in them by prior states of affairs. In this case, particles at earlier times do not *compel* later particles to act; they create a set of inclinations that pressure future particles to behave in a certain way, but it is the future particles themselves (at the moment they become present) that opt to follow those inclinations. Like infants and some nonhuman animals, it may be inevitable that particles will follow their inclinations, which accounts for their predictability. This is consistent with their actions being in a certain sense freely chosen: the particles *themselves* act rather than being compelled to act by past events.

I'm not sure whether this view is true. But it seems to me to be the view you ought to go for if you believe in free will. The other option for accommodating free will is dualism. The dualist proponent of free will carves up causation into two very different kinds: human actions—which involve free responses to rational considerations—and everything else—which acts because it's compelled by earlier events. This bifurcation is ugly and unnecessary. The simplest hypothesis consistent with human freedom is what I've just described, the view that all of nature acts freely (although not necessarily unpredictably). If it turned out that particles act out of compulsion, then my bet would be that humans do too and free will is an illusion.

Much is still unknown about the physical world and the human situation. But as yet, I can see no real grounds for denying that our choices are free. And if we are free, then it's reasonable to suppose that free will wasn't some bit of magic that popped into existence at the moment of evolutionary history when human beings first came on the scene. Human beings are embedded in, and continuous with, the rest of nature. Natural selection can only work with the resources provided by the physical world. If humans are free, then so too is the matter of which we are made.

SPIRITUALITY NATURALIZED

Immediately after fascism was defeated in 1945, Aldous Huxley—the author of the dystopian novel *Brave New World*—published a book defending what he called "The Perennial

Philosophy." The idea was that there is a single set of universal truths which are at the root of all of the major world religions. These truths, Huxley argued, can be known directly when in a state of consciousness which, although rare, has been achieved by individuals in all cultures as far back as it is possible to ascertain. Whether through gift of nature or through rigorous training, Moses, the Buddha, Jesus, the Sufi mystics of Islam, and the Vedantic mystics of Hinduism have all reached these special states of consciousness; and their insights, according to Huxley, form a common core to all religions.

Many academics have challenged Huxley's claim of religious universalism, which is difficult to sustain given that the stated beliefs of the great world religions plainly contradict each other. But the thesis that there is a distinctive kind of mystical experience that is cross-culturally pervasive is well supported. The nature of these experiences is perhaps most fully articulated in the Hindu tradition of Advaita Vedanta, which traces its roots back to the first millennium BCE but was most prominently defended by the eighth-century scholar Adi Shankara.

In ordinary states of consciousness, there is a distinction between the *subject* and the *object* of the experience, that is to say between *the thing which has the experience* (e.g., me or you) and *the things which are experienced* (e.g., pleasures, pains, sensations, etc.). But in the mystical experience, this division apparently collapses and the mystic enjoys a state of *formless consciousness.* More dramatically, mystics claim that it becomes apparent in such states that formless consciousness is the backdrop to all individual conscious experiences and hence that in a significant sense formless consciousness is the ultimate nature of each and every conscious mind. This realization allegedly undermines ordinary understanding of the distinctions between different people and leads to a conviction that in some deep sense "we are all one."[22]

Of course, just because people have these experiences, it doesn't follow that they correspond to anything real. I said in the last section that I am not certain whether or not our experiences of free will can be trusted. I am even less confident about mystical experiences. At least in the case of free will, I regularly have the experience in question. Sadly, despite daily meditation, I have not to date had the pleasure of experiencing formless consciousness.

Having said that, as in the case of free will, I am not persuaded by the arguments that have been advanced to try to show that mystical experiences must be delusions. Probably the most common reason for supposing this is the assumption that mystical experiences purport to reveal a supernatural realm: a state of formless consciousness beyond space and time that is the ultimate ground of all being. As we discussed in chapter 2, it is an important principle in both science and philosophy that we should keep our theories of reality as simple as possible. Better to hold that mystical experiences are vivid hallucinations rather than to add a supernatural entity to our theory of the world.

However, for the panpsychist there is another option. Rather than taking formless consciousness to be something *beyond the physical universe,* the panpsychist could maintain that formless consciousness is the ultimate nature of *physical reality,* or at least some aspect of it.

Here's one model of how that might work. According to the theory of physical reality known as "super-substantivalism," physical objects—tables, chairs, rocks, planets—are not distinct from spacetime, but are rather identical with *mass-instantiating regions of spacetime.*[23] That is to say, spacetime is not some great container that physical objects are located *in*. At a fundamental level, all that really exists is spacetime. But there is a distinction between those regions of spacetime that are "massy" and those that are not, and the massy ones we call "physical objects." On this view, you are nothing more than a massy person-shaped region of spacetime.

Like any purely physical theory, super-substantivalism is incomplete. Physical descriptions of mass and spacetime tell us only what these entities do and are silent on their intrinsic nature (as discussed at length in chapter 4). Suppose we fill in the view with panpsychism. We then have to find a form of consciousness to be the intrinsic nature of spacetime itself—considered in isolation from the mass that fills some of it—and we also need to find forms of consciousness to be the intrinsic nature of mass-instantiating regions of spacetime. If—and that's a big *if*—one wanted to make sense of mystical experiences as nondelusional experiences, then one could hold that:

1. Formless consciousness is the intrinsic nature of spacetime itself, in a way that is not localized but equally present at all regions of spacetime.
2. Ordinary states of consciousness are the intrinsic nature of massy regions of spacetime.

This would allow us a way of making sense of mystics' claims without committing to anything beyond the physical universe. Rather than being supernatural, formless consciousness is the intrinsic nature of a physical entity: spacetime. As the more fundamental element of each particular conscious mind, formless consciousness is the backdrop of each and every conscious experience. We can think of spacetime as providing the universal aspect of each experience, while mass and other physical properties constitute the separate and distinct contents of different experiences. Spacetime on its own is a simple and ubiquitous experience, but when combined with the forms of experience involved in complex combinations of physical properties it is transformed into a particular instance of subject/object consciousness, such as a human mind. Spacetime provides the clay and physical properties the mold.

Alternately, one might speculate that mystics are experiencing *what lies beneath spacetime*. This connects up nicely with one of the most interesting objections to panpsychism I have come across, formulated by the philosopher Susan Schneider.[24] The holy grail of modern physics is a theory of quantum gravity, a theory that can reconcile our scientific understanding of gravity with quantum mechanics. Some physicists have suggested that the way forward is to take spacetime—or at least some of its dimensions—to be built out of more fundamental stuff. Just as water is composed of stuff that is not itself water, so spacetime, on this view, is made out of stuff that is not itself spacetime.

Why should we think this is a worry for the panpsychist? Schneider says the following:

> If the more fundamental ingredients of reality are non-spatiotemporal, it is difficult to see how they can also be experiential. For if there is no time at this level, how could there be experience? Conscious experience has a felt quality that involves flow; thoughts seem to be present in the "now," and they change from moment to moment. Timeless experience is an oxymoron.[25]

It is true that most if not all of our everyday experiences involve temporal flow. But mystics speak of their experience of a timeless reality at the root of things. This is entirely in keeping with those speculative models in which spacetime does not exist at the fundamental level.

Hooking up mysticism and physics rings alarm bells for many. Even if we don't have to commit to anything supernatural, one may remain skeptical of the possibility of discovering from one's armchair—or crouched in a lotus position—fundamental truths about the nature of reality. Sam Harris has written on what he sees as the important insights of mystical experiences.[26] But such insights are, for Harris, confined to truths about the nature of our minds and can tell us nothing about the nature of the universe in general. If panpsychism is true, however, this distinction collapses, as consciousness is the intrinsic nature of physical reality. If—I emphasize the *if* one last time—the insights of mystics are correct that formless consciousness is an essential component of each and every conscious experience, then—in conjunction with panpsychism—it follows that formless consciousness is an essential component of each and every physical entity.

I am not endorsing this view. Maybe I will if I ever have a mystical experience (keep an eye on my Twitter feed if you want to be the first to know . . .). But it has a couple of cool implications.

It entails that there is, in a certain sense, life after death. My individual conscious mind will unravel and cease to be at the moment of bodily death. But one essential component of my mind—formless consciousness which is the backdrop to all of my experiences—does not cease to be. Hindus place great emphasis on meditation and good conduct as the key to realizing one's identity with formless consciousness and thus avoiding an endless cycle of pointless rebirths. But if we assume that the doctrines of karma and rebirth are false, then at death each of us collapses eternally back into formless consciousness as a matter of course. Enlightenment is guaranteed!

This view may also allow us to make sense of an objective foundation for ethics. Sam Harris and Steven Pinker are passionate believers in objective moral truth.[27] And yet, as countless commentators have observed, there doesn't seem to be anything in their worldview that can account for the reality of objective moral truth. As naturalists, they accept as real only those entities that empirical scientific investigation gives us grounds for believing in. But scientific investigation reveals what *is* the case not what *ought* to be the case. The data of observation and experiments can tell us how best to achieve our goals. But it's hard to see how an experiment or an observation could tell us what goals we ought to adopt in the first place.

Harris is adamant that science is the source of ethical knowledge, in virtue of its capacity to yield knowledge of what is most conducive to human well-being. It is of course true that empirical investigation is a good way of working out how to make people happy, but it doesn't follow that empirical investigation can shed light on the truly foundational ethical claim here, which is that human happiness or well-being is of *objective moral value.*

At this point, Harris appeals to how *uncontroversial* it is that well-being is of moral value:

> Even if each conscious being has a unique nadir on the moral landscape, we can still conceive of a state of the universe in which everyone suffers as much as he or she (or it) possibly can. If you think we cannot say this would be "bad," then I don't know what you could mean by the word "bad" (and I don't think you know what you mean

> by it either). . . . It seems uncontroversial to say that a change that leaves everyone worse off, by any rational standard, can be reasonably called "bad," if this word is to have any meaning at all.[28]

It's not clear how this is supposed to help us to make sense of the foundations of ethical truth. Let us agree with Harris that maximal and universal suffering would be an objectively bad thing. It is one thing to *accept* this moral truth; it is quite another thing to give an *explanation* of what makes it true. Here's an analogy. Everyone agrees that things fall to the ground. What a scientist tries to work out is *why* this happens. We would not be impressed with Newton or Einstein if they had simply vigorously asserted "Who would deny that apples fall to the ground???" The point is not to assert what everyone already accepts but to explain it. Similarly, in the ethical case we want to know *why* suffering is objectively bad; what is it about reality that makes this the case?

At some point one reaches bedrock; explanation has to come to an end somewhere. Einstein's explanation of gravity ends with general facts about the reciprocal causal relationship between matter and spacetime. But Harris's account of moral truth doesn't even begin. According to commonsense morality, there is something *objectively wrong* with someone who tortures children for fun. Presumably, Harris would wholeheartedly agree. But what he does not provide, which is what we really want, is an account of what makes it the case that there is something objectively wrong with the sadist. What is it about reality that grounds this truth?

The Christian apologist William Lane Craig has forcefully pressed this point on Harris, arguing that the only solution is to postulate a God whose commands can serve as the grounds of objective moral truth.[29] This theory has the disadvantage of being both profligate and unsatisfying. For we now have the question: What makes it the case that God's commands have moral force? One might argue that it is an obvious truth of morality that if a perfectly good being commands something, then we ought to do it. Perhaps so. It is also an obvious moral truth, as Harris is keen to point out, that if someone is in pain, then you ought to help them. But what we are looking for is an explanation of *why* obvious moral truths are true. I can't see how stopping with the obvious moral truth (if indeed it is one) that we should do what God tells us is any more satisfying than stopping with the obvious moral truth that pain is bad.

The view of the mystics, in contrast, does provide a satisfying account of the objectivity of ethics. Selfish conduct is rooted in the belief that we are wholly separate and distinct individuals. The sadist enjoys another's pain only if she is not suffering *herself.* But according to the mystics, this belief in the total separateness of people is false. There are distinct conscious minds, but they are not *wholly* distinct; your mind and my mind overlap. Indeed, the most basic element of *my* mind—the formless consciousness which forms the backdrop of each experience—is identical with the most basic element of *your* mind. Your consciousness and my consciousness are, as it were, painted on one and the same canvas. According to the testimony of mystics, it is this realization that results in the boundless compassion of the enlightened. If the mystical hypothesis is true, then ethical objectivity is grounded in the nature of reality. The sadist is objectively flawed for exactly the reason that the flat earther is: both have a false view of reality.

If we were to adopt the panpsychist interpretation of the mystical hypothesis, we could secure these advantages with zero cost, as this theory commits us to nothing beyond the physical universe we all believe in anyway. True, it offers a specific account of the intrinsic nature of physical reality, but physical reality has to have some intrinsic nature or other, and I can't see any reason to think that the proposal under consideration is more profligate than any other.

I am still hesitant about endorsing this view. Its implications have profound significance for the meaning of human life. But we should be trying to work out which view is most likely to be true not which view we would most like to be true. While I have access to the datum that motivates panpsychism, namely conscious experience, I do not have access to the putative evidence that would make it rational for me to accept the mystical worldview.

The American psychologist William James got it about right at the end of his classic analysis of mystical experience in *The Varieties of Religious Experience.* Most of the discussion is taken up with a study of the experiences themselves, but toward the end he turns to the question of whether the existence of these experiences can give us reason to accept the mystical view. James's conclusion is balanced. On the one hand:

> As a matter of psychological fact, mystical states of a well-pronounced and emphatic sort *are* usually authoritative over those who have them. They have been "there," and know. It is vain for rationalism to grumble about this. . . . Our own more "rational" beliefs are based on evidence exactly similar in nature to that which mystics quote for theirs. Our senses, namely, have assured us of certain states of fact; but mystical experiences are as direct perceptions of fact for those who have them as any sensations ever were for us.[30]

There is no way the mystic can *prove* that their experiences correspond to reality. But nor can we prove that our sensory experiences correspond to reality. As philosophers know only too well, there is no way of ruling out that we are in the Matrix enjoying illusory experiences created in us by computers. We simply have to take it for granted that our senses tell us the truth. If it's okay for us to trust our sensory experiences, how could it be irrational for mystics to do the same with respect to their mystical experiences? To accuse the mystic of irrationality would be to employ a double standard.

On the other hand:

> . . . mystics have no right to claim that we ought to accept the deliverance of their peculiar experiences, if we are ourselves outsiders and feel no private call thereto. The utmost they can ever ask of us in this life is to admit that they establish a presumption. They form a consensus and have an unequivocal outcome; and it would be odd, mystics might say, if such a unanimous type of experience should prove to be altogether wrong. At bottom, however, this would only be an appeal to numbers, like the appeal of rationalism the other way; and the appeal to numbers has no logical force. If we acknowledged it, it is for "suggestive," not for logical reasons: we follow the majority because to do so suits our life.[31]

The appropriate attitude to mystical experiences for those who haven't had them is probably one of agnosticism, the withholding of belief either that mystical experiences provide genuine insight into the nature of reality or that they are delusions. Having said that, I can't help being excited by the possibility that, in a panpsychist worldview, the yearnings of faith and the rationality of science might finally come into harmony. Perhaps it's okay to hope. And to keep meditating.

A UNIVERSE OF MEANING

In the early twentieth century, Max Weber wrote of the "disenchantment" of nature caused by modernity and the rise of capitalism. In a religious or traditional worldview, the universe is filled with meaning and purpose; as Weber put it, "the world remained a great magical garden."[32] The modern scientific worldview, in contrast, seems to present us with an immense universe entirely devoid of meaning, in which human beings are a tiny and painfully temporary accident.

This can lead to a sense of alienation. We seem to have nothing in common with the universe, no real home within it. The "big picture" story of the universe is one of insentient and meaningless physical processes, from which we are a senseless aberration. In the absence of a place in the universe, we have only consumerism and the endless quest for economic growth to make sense of our lives.

This problem of "cosmic alienation" has grown worse as society has become more globalized. When one is embedded in a traditional society, ignorant of the plurality of social forms across the globe, the conditioned meanings of one's society seem to define the cosmos. One lives not in a meaningless universe but in a world with sense and purpose. However, globalized markets have eroded many traditional forms of life; international chain stores have conquered the centers of communities; advertising now fills all corners of public space. Where local beauty is preserved, it is only as a quaint museum piece for globe-trotting tourists.

Even the mere awareness of a plurality of cultural forms can lead to alienation, by making it plain that one's own social and moral norms are not timeless realities but contingent choices of history. Traditional ways of life come to be seen as empty of meaning, leading to relativism or even nihilism. It is perhaps no surprise that nationalism is once again on the rise, as people grasp after something probably lost forever. Without the meaningful structures once given by traditional society, we are left with nothing but mechanistic nature and the meaningless abyss of empty space.

Panpsychism offers a way of "re-enchanting" the universe. On the panpsychist view, the universe is *like us;* we *belong* in it. We need not live exclusively in the human realm, ever more diluted by globalization and consumerist capitalism. We can live in nature, in the universe. We can let go of nation and tribe, happy in the knowledge that there is a universe that welcomes us. My hope is that panpsychism can help humans once again to feel that they have a place in the universe. At home in the cosmos, we might begin to dream about—and perhaps make real—a better world.

Notes

1. *How Galileo Created the Problem of Consciousness*

1. Seth, "The Real Problem."
2. Paley, *Natural Theology.*
3. Dawkins, *The Blind Watchmaker.*
4. Paul Churchland, *Matter and Consciousness.*
5. This was the start of Minkowski's address to the 80th Assembly of German Natural Scientists and Physicians in September 21, 1908, as recorded in his "Space and Time," 72.
6. Galileo, *The Assayer,* 237–38.

2. *Is There a Ghost in the Machine?*

1. Bloom, *Descartes' Baby.*
2. Hume, *An Enquiry Concerning Human Understanding,* VII.
3. Chalmers, "Facing Up to the Problem of Consciousness"; Chalmers, *The Conscious Mind.*
4. London and Bauer, "The Theory of Observation in Quantum Mechanics"; Wigner, "Remarks on the Mind-Body Question," 169.
5. Stapp, *Mind, Matter and Quantum Mechanics.*
6. Chalmers and McQueen, "Consciousness and the Collapse of the Wave Function"; McQueen's article "Does Consciousness Cause Quantum Collapse?" was an important source in writing this section.
7. von Neumann, *Mathematical Foundations of Quantum Theory.*
8. Everett, " 'Relative State' Formulation of Quantum Mechanics."
9. Einstein, "On the Method of Theoretical Physics."

3. *Can Physical Science Explain Consciousness?*

1. Krauss, "The Consolation of Philosophy."
2. Ladyman and Ross, *Every Thing Must Go*, 29.
3. Patricia S. Churchland, *Touching a Nerve*, 60.
4. This thought experiment was outlined in Galileo's unpublished book *De Moto* ("On Motion").
5. Nagel, *The View from Nowhere.*
6. Ludlow, Nagasawa, and Stoljar, *There's Something About Mary.*
7. Leibniz, *The Monadology*, Section 17.
8. Jackson, "Epiphenomenal qualia." In the same year, the philosopher Howard Robinson published a very similar thought experiment pressing essentially the same argument in his book *Matter and Sense.* Robinson's argument is rather unfairly much less known, perhaps because the missing knowledge in his thought experiment was of the experience of sounds rather than of colors.
9. Dennett, *Consciousness Explained;* Paul M. Churchland, "Consciousness and the Introspection of 'Qualitative Simples.'"
10. Dennett, *Consciousness Explained*, 400.
11. Nordby, "What Is This Thing You Call Color?," 79–82.
12. Nordby, "What Is This Thing You Call Color?," 77.
13. Dennett, *Consciousness Explained*, 399–400.
14. This use of the term "zombie" was coined by Robert Kirk in his papers "Zombies vs. Materialists" and "Sentience and Behaviour." David Chalmers gave a very detailed and influential defense of the zombie argument in *The Conscious Mind.*
15. Seth, "Conscious Spoons, Really? Pushing Back Against Panpsychism." Although the title of this blog post sounds very hostile to panpsychism, Seth is not hostile to the *philosophical case* for panpsychism (personal communication). The blog post responds to an article in *Quartz* magazine, which was based on interviews with me, David Chalmers, and Hedda Hassel Mørch. All three of us tried to emphasize in the interview that the vast majority of panpsychists would deny that spoons are conscious.
16. Frankish, "Illusionism as a Theory of Consciousness," 12–13.
17. Strawson, "The Consciousness Deniers."
18. Searle, "Minds, Brains and Programs."
19. Humphrey, *Soul Dust.*
20. Levine, *Purple Haze.*

4. *How to Solve the Problem of Consciousness*

1. This comparison is made by David Chalmers in his classic article "Facing Up to the Problem of Consciousness" (in which he coined

the phrase "the hard problem"), although Chalmers uses it to defend dualism rather than panpsychism.

2. Strawson, "Realistic Materialism: Why Physicalism Entails Panpsychism," p. 29.
3. Quoted in Isaacson, *Einstein: His Life and Universe,* p. 262.
4. Eddington, *The Nature of the Physical World,* Chapter 12.
5. Taken from a survey conducted by PhilPapers, available at https://philpapers.org.
6. Eddington, *The Nature of the Physical World,* Chapter 12.
7. Hawking, *A Brief History of Time,* 174.
8. Eddington, *The Nature of the Physical World,* Chapter 12.
9. Eddington, *The Nature of the Physical World,* Chapter 13.
10. Eddington, *The Nature of the Physical World,* Chapter 12.
11. Blackmore, "First Person—Into the Unknown."
12. Episode 25 of *The Panpsycast,* available at http://thepanpsycast.com.
13. Papineau, "The Problem of Consciousness."
14. Papineau, "The Problem of Consciousness."
15. James, *Principles of Psychology,* 160. Although the problem itself goes back to James, the term "combination problem" comes from William Seager's article "Consciousness, Information, and Panpsychism."
16. James, *Principles of Psychology,* 160.
17. Descartes, *Meditations on First Philosophy,* 59.
18. Sperry, "Consciousness, Personal Identity, and the Divided Brain."
19. Gazzaniga sums up many of his findings about the division of labor between the two hemispheres in "Cerebral Specialization and Interhemispheric Communication."
20. Roelofs, *Combining Minds.*
21. Laplace, *A Philosophical Essay on Probabilities,* 4.
22. Hendry, *The Metaphysics of Chemistry.*
23. Mørch, *Panpsychism and Causation.*
24. Mørch, "The Integrated Information Theory of Consciousness."
25. Teilhard de Chardin, *The Phenomenon of Man.*
26. Russell, *The Analysis of Matter,* 325.

5. *Consciousness and the Meaning of Life*

1. Zorgon, *The Hitchhiker's Guide to the Many Realities.*
2. Watts et al., "The *Lancet* Countdown on Health and Climate Change."
3. Holder, Kommenda, and Watts, "The Three-Degree World."
4. A good source on this is https://www.skepticalscience.com/.
5. Hume, *An Enquiry Concerning Human Understanding,* XII: 23.
6. Hume, *An Enquiry Concerning Human Understanding,* XII: 24.

7. The relevance of Hume for our times has also been noted by Julian Baggini in his article "Hume the Humane."
8. Based on a poll in 2017 by ComRes, commissioned by the Energy and Climate Intelligible Unit, available at https://www.comresglobal.com.
9. https://climateactiontracker.org.
10. Klein, *This Changes Everything*, 177.
11. Klein, *This Changes Everything*, 170.
12. Papineau, "The Problem of Consciousness."
13. Gagliano et al., "Learning by Association in Plants."
14. Quotation from "Is Eating Plants Wrong?," BBC Radio 4. Available as a podcast at https://www.bbc.co.uk.
15. Novoplansky et al., "Plant Responsiveness to Root–Root Communication of Stress Cues."
16. Simard, "Mycorrhizal Networks Facilitate Tree Communication, Learning and Memory."
17. van Inwagan, *An Essay on Free Will*, Chapter 4.
18. Lowe, *Personal Agency*, Part II.
19. Libet et al., "Time of Conscious Intention to Act in Relation to Onset of Cerebral Activity (Readiness Potential)"; Libet, "Unconscious Cerebral Initiative and the Role of Conscious Will in Voluntary Action."
20. Dennett, *Freedom Evolves*, Chapter 8.
21. Soon et al., "Unconscious Determinants of Free Decisions in the Human Brain."
22. Miri Albahari has explored and defended this kind of "Advaitic" view, for example in her article "Beyond Cosmopsychism and the Great I Am: How the World Might Be Grounded in Universal 'Advaitic' Consciousness." My description of the view is very much influenced by her work.
23. Schaffer, "Spacetime: The One Substance."
24. Schneider, "Spacetime Emergence, Panpsychism and the Nature of Consciousness."
25. Schneider, "Spacetime Emergence, Panpsychism and the Nature of Consciousness."
26. Harris, *Waking Up*.
27. Harris, *The Moral Landscape;* Pinker, *Enlightenment Now*.
28. Harris, *The Moral Landscape*, 39.
29. Craig and Harris debated at the University of Notre Dame, Indiana, April 2011. Available at https://samharris.org and https://www.reasonablefaith.org.
30. James, *The Varieties of Religious Experience*, Lectures XVI and XVII.
31. James, *The Varieties of Religious Experience*, Lectures XVI and XVII.
32. Weber, *The Sociology of Religion*.

Bibliography

Albahari, Miri. "Beyond Cosmopsychism and the Great I Am: How the World Might Be Grounded in Advaitic Consciousness," in W. Seager, ed., *The Routledge Handbook of Panpsychism,* London, New York: Routledge, forthcoming.

Baggini, Julian. "Hume the Humane." *Aeon,* August 15, 2018, https://aeon.co.

Blackmore, Susan. "First Person—Into the Unknown." *New Scientist,* November 4, 2000.

Bloom, Paul. *Descartes' Baby: How the Science of Child Development Explains What Makes Us Human.* New York: Basic Books, 2004.

Cartwright, Nancy. *How the Laws of Physics Lie.* New York: Oxford University Press, 1983.

Chalmers, David. *The Conscious Mind.* New York: Oxford University Press, 1996.

———. "Facing Up to the Problem of Consciousness." *Journal of Consciousness Studies* 2, no. 3 (2005): 200–19.

Chalmers, David, and McQueen, Kelvin. "Consciousness and the Collapse of the Wave Function," in S. Gao, ed., *Quantum Mechanics and Consciousness.* New York: Oxford University Press, forthcoming.

Churchland, Patricia S. *Touching a Nerve.* New York: W. W. Norton, 2013.

Churchland, Paul, M. "Consciousness and the Introspection of 'Qualitative Simples,'" in R. Brown, ed., *Consciousness Inside and Out: Phenomenology, Neuroscience, and the Nature of Experience.* Dordrecht/Heidelberg/New York/London: Springer, 2013.

———. *Matter and Consciousness,* revised edition. Cambridge: MIT Press, 1988.

Coleman, Sam. "Panpsychism and Neutral Monism: How to Make Up

One's Mind," in G. Brüntrup and L. Jaskolla, eds., *Panpsychism: Contemporary Perspectives.* New York: Oxford University Press, 2016.
———. "The Real Combination Problem: Panpsychism, Micro-Subjects and Emergence." *Erkenntnis* 79, no. 1 (2014): 19–44.
Dawkins, Richard. *The Blind Watchmaker: Why the Evidence of Evolution Reveals a Universe Without Design.* New York: W. W. Norton, 1986.
Dennett, Daniel C. *Consciousness Explained.* Boston: Little, Brown, 1991.
———. *Freedom Evolves.* New York: Viking, 2003.
Descartes, René. *Meditations on First Philosophy,* in *Meditations on First Philosophy with Selections from the Objections and Replies,* edited and translated by J. Cottingham. Cambridge: Cambridge University Press, 1996; originally published in 1645.
Eddington, Arthur Stanley. *The Nature of the Physical World.* London: Macmillan, 1928.
Einstein, Albert. "On the Method of Theoretical Physics." *The Herbert Spencer Lecture,* delivered at Oxford, June 10, 1933.
Everett, Hugh. "Relative State Formulation of Quantum Mechanics." *Review of Modern Physics,* 29 (1957): 454–62.
Frankish, Keith. "Illusionism as a Theory of Consciousness." *Journal of Consciousness Studies,* 23 (2016): 11–12.
Gagliano, Monica, et al. "Learning by Association in Plants." *Scientific Reports* 6, article no. 38427 (2016).
Galileo Galilei. *The Assayer,* originally published in 1623, in *Discovering and Opinions of Galileo,* edited by Stillman Drake. New York: Anchor, 1957.
Gazzaniga, Michael S. "Cerebral Specialization and Interhemispheric Communication. Does the Corpus Callosum Enable the Human Condition?" *Brain* 123, no. 7 (2000): 1293–336.
Goff, Philip. *Consciousness and Fundamental Reality.* New York: Oxford University Press, 2017.
———. "Essentialist Modal Rationalism." *Synthese,* forthcoming.
Goff, Philip, et al. "Panpsychism," in E. N. Zalta, ed., *The Stanford Encyclopedia of Philosophy,* 2017, https://plato.stanford.edu.
Harris, Sam. *The Moral Landscape: How Science Can Determine Human Values.* New York: Free Press, 2010.
———. *Waking Up: A Guide to Spirituality Without Religion.* New York: Simon & Schuster, 2014.
Hawking, Stephen. *A Brief History of Time: From the Big Bang to Black Holes.* New York: Bantam, 1988.
Hawking, Stephen, and Leonard Mlodnow. *The Grand Design.* New York: Bantam, 2010.
Hendry, Robin. *The Metaphysics of Chemistry.* Oxford: Oxford University Press, forthcoming.

Holder, Josh, Niko Kommenda, and Jonathan Watts. "The Three Degree World: The Cities That Will Be Drowned by Global Warming." *The Guardian*, November 3, 2017.

Hume, David. *An Enquiry Concerning Human Understanding*, edited with an introduction and notes by Peter Millican. Oxford: Oxford University Press, 2007; originally published in 1748.

Humphrey, Nicholas. *Soul Dust: The Magic of Consciousness*. Princeton: Princeton University Press, 2011.

Huxley, Aldous. *The Perennial Philosophy*. New York: Harper & Brothers, 1945.

Isaacson, Walter. *Einstein: His Life and Universe*. New York: Simon & Schuster, 2007.

Jackson, Frank. "Epiphenomenal qualia." *The Philosophical Quarterly* 32, no. 127 (1982): 127–36.

James, William. *Principles of Psychology*, vol. I. Cambridge: Harvard University Press, 1981; originally published in 1890.

———. *The Varieties of Religious Experience: A Study in Human Nature*. London and Bombay: Longmans, Green & Co., 1902.

Kirk, Robert. "Sentience and Behavior." *Mind* 83, no. 329 (1974): 43–60.

———. "Zombies vs. Materialists." *Proceedings of the Aristotelian Society* 48, (1974): 135–63.

Klein, Naomi. *This Changes Everything: Capitalism vs. the Climate*. New York: Simon & Schuster, 2014.

Kraus, Lawrence, M. "The Consolation of Philosophy." *Scientific American*, April 27, 2012.

Kripke, Saul. *Naming and Necessity*. Cambridge: Harvard University Press, 1990.

Ladyman, James, and Don Ross (with David Spurrett and John Collier). *Every Thing Must Go*. Oxford: Oxford University Press, 2007.

Laplace, Pierre Simon. *A Philosophical Essay on Probabilities*, translated by F. W. Truscott and F. L. Emory. Mineola, NY: Dover Publications, 1951.

Leibniz, G. W. *The Monadology*, in *G. W. Leibniz: Philosophical Texts*, edited and translated by R. S. Woolhouse and R. Francks, Oxford: Oxford University Press, 1998; originally published in 1714.

Levine, Joseph. *Purple Haze: The Puzzle of Consciousness*. Oxford and New York: Oxford University Press, 2004.

Libet, Benjamin, W. "Unconscious Cerebral Initiative and the Role of Conscious Will in Voluntary Action." *Behavioral Brain Sciences* 8 (1985): 529–66.

Libet, Benjamin W., et al. "Time of Conscious Intention to Act in Relation to Onset of Cerebral Activity (Readiness Potential): The Unconscious Initiation of a Freely Voluntary Act." *Brain* 106, no. 3 (1983): 623–42, 9.

Locke, John. *An Essay Concerning Human Understanding.* Edited by Pauline Phemister. Oxford: Clarendon Press, 2008; originally published in 1689.

Lodge, David. *Thinks* . . . London: Secker & Warburg, 2001.

London, Fritz, and Edmond Bauer. "The Theory of Observation in Quantum Mechanics," in J. A. Wheeler and W. H. Zurek, eds., *Quantum Theory and Measurement.* Princeton: Princeton University Press, 1983.

Lowe, E. J. *Personal Agency: The Metaphysics of Mind and Action.* Oxford: Oxford University Press, 2008.

Ludlow, Peter, Yujin Nagasawa, and Daniel Stoljar, eds. *There's Something About Mary: Essays on Phenomenal Consciousness and Frank Jackson's Knowledge Argument.* Cambridge: MIT Press, 2004.

McQueen, Kelvin. "Does Consciousness Cause Quantum Collapse?" *Philosophy Now,* August/September 2017.

Minkowski, Hermann. "Space and Time" in Hendrik A. Lorentz, Albert Einstein, Hermann Minkowski, and Hermann Weyl, eds., *The Principle of Relativity: A Collection of Original Memoirs on the Special and General Theory of Relativity.* New York: Dover (1952): 75–91.

Mørch, Hedda Hassel. "The Integrated Information Theory of Consciousness." *Philosophy Now,* August/September 2017.

———. "Panpsychism and Causation: A New Argument and a Solution to the Combination Problem." PhD diss., University of Oslo, 2014.

Nagel, Thomas. *The View from Nowhere.* New York: Oxford University Press, 1986.

———. "What Is It Like to Be a Bat?" *Philosophical Review* 83, no. 4 (1974): 435–50.

Nordby, Knut. "What Is This Thing You Call Color? Can a Totally Color-Blind Person Know About Color?," in T. Alter and S. Walter, eds., *Phenomenal Concepts and Phenomenal Knowledge: New Essays on Phenomenal Concepts and Physicalism.* New York: Oxford University Press, 2007.

Novoplansky, Ariel, et al. "Plant Responsiveness to Root–Root Communication of Stress Cues." *Annals of Botany* 110, no. 2 (2012): 271–80.

Paley, William. *Natural Theology or Evidence for the Existence and Attributes of the Deity, Collected from the Appearances of Nature,* edited with an introduction and notes by M. D. Eddy and D. Knight. Oxford and New York: Oxford University Press, 2006; originally published in 1809.

Papineau, David. "The Problem of Consciousness," in U. Kriegel. ed., *The Oxford Handbook of Consciousness.* Oxford: Oxford University Press, forthcoming.

Pinker, Steven. *Enlightenment Now: The Case for Reason, Science, Humanism, and Progress.* New York: Viking, 2018.

Robinson, Howard. *Matter and Sense: A Critique of Contemporary Materialism.* Cambridge: Cambridge University Press, 1982.

Roelofs, Luke. *Combining Minds: How to Think About Composite Subjectivity.* New York: Oxford University Press, 2019.

Russell, Bertrand. *The Analysis of Matter.* London: Kegan Paul, 1927.

Schaffer, Jonathan. "Spacetime: The One Substance." *Philosophical Studies* 145, no. 1 (2009): 131–48.

Schneider, Susan. "Spacetime Emergence, Panpsychism and the Nature of Consciousness." *Scientific American,* August 6, 2018.

Seager, William. "Consciousness, Information, and Panpsychism." *Journal of Consciousness Studies* 2, no. 3 (1995): 272–88.

Searle, J. "Minds, Brains and Programs." *Behavioral and Brain Sciences* 3, no. 3 (1980): 417–57.

Seth, Anil K. "Conscious Spoons, Really? Pushing Back Against Panpsychism." *NeuroBanter* (blog), February 1, 2018.

———. "The Real Problem." *Aeon,* 2016, https://aeon.co/.

Simard, Suzanne, W. "Mycorrhizal Networks Facilitate Tree Communication, Learning and Memory," in F. Baluska, M. Gagliano, and G. Witzany, eds., *Memory and Learning in Plants.* Dordrecht/Heidelberg /New York/London: Springer, 2018.

Skrbina, David. *Panpsychism in the West.* Cambridge: MIT Press, 2007.

Soon, Chun Siong, et al. "Unconscious Determinants of Free Decisions in the Human Brain." *Nature Neuroscience* 11 (2008): 543–45.

Sperry, Roger. "Consciousness, Personal Identity, and the Divided Brain." *Neuropsychologia,* 22, no. 6 (1984): 661–73.

Stapp, Henry P. *Mind, Matter and Quantum Mechanics.* Berlin/Heidelberg: Springer Verlag, 2009.

Strawson, Galen. "The Consciousness Deniers." *New York Review of Books,* March 13, 2018.

Strawson, Galen. "Realistic Materialism: Why Physicalism Entails Panpsychism." *Journal of Consciousness Studies* 13, nos. 10–11 (2006): 3–31.

Teilhard de Chardin, Pierre. *The Phenomenon of Man,* translated by B. Wall. New York: Harper & Brothers, 1959.

Turing, Alan M. "Computing Machinery and Intelligence." *Mind* 59, no. 236 (1950): 433–60.

van Helden, Albert. "On Motion." *The Galileo Project,* 1995, https://galileo.rice.edu.

van Inwagan, Peter. *An Essay on Free Will.* Oxford: Oxford University Press, 1983.

von Neumann, John. *Mathematical Foundations of Quantum Theory.* Berlin: Julius Springer, 1932.

Watts, Nick, et al. "The *Lancet* Countdown on Health and Climate Change:

From 25 Years of Inaction to a Global Transformation for Public Health." *The Lancet* 391, no. 10120 (2018): 581–630, https://www.thelancet.com.

Weber, Max. *The Sociology of Religion.* Boston: Beacon Press, 1993; first published in 1920.

Wigner, Eugene. "Remarks on the Mind-Body Question," in J. A. Wheeler and W. H. Zurek, eds., *Quantum Theory and Measurement.* Princeton: Princeton University Press, 1983.

Zorgon, E. T. *The Hitchhiker's Guide to the Many Realities.* Interdimensional Inc.

Index